Lecture Notes in Computer Science   12929

More information about this subseries at http://www.springer.com/series/7412

Mauricio Reyes · Pedro Henriques Abreu ·
Jaime Cardoso · Mustafa Hajij · Ghada Zamzmi ·
Paul Rahul · Lokendra Thakur (Eds.)

Interpretability of Machine Intelligence in Medical Image Computing, and Topological Data Analysis and Its Applications for Medical Data

4th International Workshop, iMIMIC 2021
and 1st International Workshop, TDA4MedicalData 2021
Held in Conjunction with MICCAI 2021
Strasbourg, France, September 27, 2021
Proceedings

Springer

Editors
Mauricio Reyes 🆔
University of Bern
Bern, Switzerland

Jaime Cardoso 🆔
INESC, FEUP
Porto, Portugal

Ghada Zamzmi
National Institutes of Health
Bethesda, MD, USA

Lokendra Thakur
Broad Institute of MIT and Harvard
Cambridge, MA, USA

Pedro Henriques Abreu 🆔
CISUC, FCTUC
Coimbra, Portugal

Mustafa Hajij
Santa Clara University
Santa Clara, CA, USA

Paul Rahul
Massachusetts General Hospital, Harvard
Boston, MA, USA

ISSN 0302-9743   ISSN 1611-3349 (electronic)
Lecture Notes in Computer Science
ISBN 978-3-030-87443-8   ISBN 978-3-030-87444-5 (eBook)
https://doi.org/10.1007/978-3-030-87444-5

LNCS Sublibrary: SL6 – Image Processing, Computer Vision, Pattern Recognition, and Graphics

This Springer imprint is published by the registered company Springer Nature Switzerland AG
The registered company address is: Gewerbestrasse 11, 6330 Cham, Switzerland

iMIMIC 2021

It is our genuine honor and great pleasure to welcome you to the 4th Workshop on Interpretability of Machine Intelligence in Medical Image Computing (iMIMIC 2021), a satellite event at the 24th International Conference on Medical Image Computing and Computer Assisted Intervention (MICCAI 2021). Following in the footsteps of the three previous successful meetings in Granada, Spain (2018), Shenzhen, China (2019), and Lima, Peru (2020), we gathered for this new edition. iMIMIC is a single-track, half-day workshop consisting of high-quality, previously unpublished papers, presented either orally, and intended to act as a forum for research groups, engineers, and practitioners to present recent algorithmic developments, new results, and promising future directions in interpretability of machine intelligence in medical image computing. Machine learning systems are achieving remarkable performances at the cost of increased complexity. Hence, they become less interpretable, which may cause distrust, potentially limiting clinical acceptance. As these systems are pervasively being introduced to critical domains, such as medical image computing and computer assisted intervention, it becomes imperative to develop methodologies allowing insight into their decision making. Such methodologies would help physicians to decide whether they should follow and trust automatic decisions. Additionally, interpretable machine learning methods could facilitate defining the legal framework of their clinical deployment. Ultimately, interpretability is closely related to AI safety in healthcare.

This year's iMIMIC was held on September 27, 2021, virtually in Strasbourg, France. There was a very positive response to the call for papers for iMIMIC 2021. We received 12 full papers and 7 were accepted for presentation at the workshop, where each paper was reviewed by at least three reviewers. The accepted papers present fresh ideas of interpretability in settings such as regression, multiple instance learning, weakly supervised learning, local annotations, classifier re-training, and model pruning. The high quality of the scientific program of iMIMIC 2021 was due to, first, to the authors who submitted excellent contributions and, second, the dedicated collaboration of the international Program Committee and the other researchers who reviewed the papers. We would like to thank all the authors for submitting their contributions and for sharing their research activities.

We are particularly indebted to the Program Committee members and to all the reviewers for their precious evaluations, which permitted us to set up this publication.

We were also very pleased to benefit from the participation of the invited speakers Mihaela van der Schaar, Cambridge University, USA: Been Kim, Google Brain: and Cynthia Rudin, Duke University, USA. We would like to express our sincere gratitude to these world-renowned experts.

September 2021

Mauricio Reyes
Pedro H. Abreu
Jaime Cardoso

Organization

General Chairs

Mauricio Reyes	University of Bern, Switzerland
Jaime Cardoso	INESC Porto, Universidade do Porto, Portugal
Himabindu Lakkaraju	Harvard University, USA
Jayashree Kalpathy-Cramer	Massachusetts General Hospital, Harvard University, USA
Nguyen Le Minh	Japan Advanced Institute of Science and Technology, Japan
Pedro H. Abreu	CISUC and University of Coimbra, Portugal
Roland Wiest	University Hospital Bern, Switzerland
José Amorim	CISUC and University of Coimbra, Portugal
Wilson Silva	INESC TEC and University of Porto, Portugal

Program Committee

Adam Perer	Carnegie Mellon University, USA
Alexander Binder	University of Oslo, Norway
Ben Glocker	Imperial College, UK
Bettina Finzel	University of Bamberg, Germany
Bjoern Menze	Technical University of Munich, Germany
Carlos A. Silva	University of Minho, Portugal
Christoph Molnar	Ludwig Maximilian University of Munich, Germany
Claes Nøhr Ladefoged	Rigshospitalet, Denmark
Dwarikanath Mahapatra	Inception Institute of AI, UAE
Ender Konukoglu	ETH Zurich, Switzerland
George Panoutsos	University of Sheffield, UK
Henning Müller	HES-SO Valais-Wallis, Switzerland
Hrvoje Bogunovic	Medical University of Vienna, Austria
Isabel Rio-Torto	University of Porto, Portugal
Islem Rekik	Istanbul Technical University, Turkey
Mara Graziani	HES-SO Valais-Wallis, Switzerland
Nick Pawlowski	Imperial College London, UK
Sérgio Pereira	Lunit, South Korea
Ute Schmid	University of Bamberg, Germany
Wojciech Samek	Fraunhofer HHI, Germany

Sponsors

Neosoma Inc.
Varian, a Siemens Healthineers Company

TDA4MedicalData 2021

TDA4MedicalData 2021 is the First International Workshop on Topological Data Analysis and its Applications for Medical Data. TDA4MedicalData 2021 proceedings contain 5 high-quality papers of 8 pages that were selected through a rigorous peer review process.

Recent years have witnessed an increasing interest in the role topology plays in machine learning and data science. Topology offers a collection of techniques and tools that have matured to a field known today as Topological Data Analysis (TDA). TDA provides a general and multi-purpose set of robust tools that have shown excellent performance in several real-world applications. These tools are naturally applicable to numerous types of data, including images, points cloud, graphs, meshes, time-varying data, and more. TDA techniques have been increasingly used with other techniques, such as deep learning, to increase the performance, and generalizability of a generic learning task. Further, the properties of the topological tools allow the discovery of complex relationships and separaton of signals that are hidden in the data from noise. Finally, topological methods naturally lend themselves to visualization, rendering them useful for tasks that require interpretability and explainability.

All these properties of topological-based methods strongly motivate the adoption of TDA tools to various applications and domains including neuroscience, bioscience, biomedicine, and medical imaging. This workshop will focus on using TDA techniques to enhance the performance, generalizability, efficiency, and explainability of the current methods applied to medical data. In particular, the workshop will focus on using TDA tools solely or combined with other computational techniques (e.g., feature engineering and deep learning) to analyze medical data including images/videos, sounds, physiological, texts and sequence data. The combination of TDA and other computational approaches is more effective in summarizing, analyzing, quantifying, and visualizing complex medical data. This workshop brought together mathematicians, biomedical engineers, computer scientists, and medical doctors for the purpose of showing the strength of using TDA-based tools for medical data analysis.

The proceedings of the workshop are published as a joint LNCS volume alongside other satellite events organized in conjunction with MICCAI. In addition to the papers, abstracts, slides, and posters presented during the workshop will be made publicly available on the TDA4MedicalData website.

We would like to thank all the speakers and authors for joining our workshop, the Program Committee for their excellent work with the peer reviews, the workshop chairs and editors for their help with the organization of the first TDA4MedicalData workshop.

September 2021

Mustafa Hajij
Ghada Zamzmi
Paul Rahul
Lokendra Thakur

Organization

General Chair

Mustafa Hajij — Santa Clara University, USA

Program Committee Chairs

Ghada Zamzmi — National Institutes of Health, USA
Rahul Paul — Massachusetts General Hospital, Harvard Medical School, USA
Lokendra Thakur — Broad Institute of MIT and Harvard, USA

Program Committee

Paul Bendich	Duke University, USA
Sema Candemir	Ohio State University, USA
Gunnar Carlsson	Stanford University, USA
Stefania Ebli	École Polytechnique Fédérale de Lausanne, Switzerland
Andrew King	King's College, UK
Nina Miolane	University of California, Santa Barbara, USA
Vic Patrangenaru	Florida State University, USA
Raul Rabadan	Columbia University, USA
Karthikeyan Natesan	Ramamurthy IBM, USA
Bastian Rieck	ETH Zurich, Switzerland
Paul Rosen	University of South Florida, USA
Aldo Guzman	Saenz IBM, USA
Manish Saggar	Stanford University, USA
Md Sirajus Salekin	University of South Florida, USA
Rajaram Sivaramakrishnan	National Institute of Health, USA
Clifford Smyth	University of North Carolina, USA
Md Taufeeq Uddin	University of South Florida, USA
Mo Zhou	Stanford University, USA

Contents

iMIMIC 2021 Workshop

Interpretable Deep Learning for Surgical Tool Management

Mark Rodrigues[1]($\boxtimes$), Michael Mayo[1], and Panos Patros[2]

[1] Department of Computer Science, University of Waikato, Hamilton, New Zealand
[2] Department of Software Engineering, University of Waikato,
Hamilton, New Zealand

Abstract. This paper presents a novel convolutional neural network framework for multi-level classification of surgical tools. Our classifications are obtained from multiple levels of the model, and high accuracy is obtained by adjusting the depth of layers selected for predictions. Our framework enhances the interpretability of the overall predictions by providing a comprehensive set of classifications for each tool. This allows users to make rational decisions about whether to trust the model based on multiple pieces of information, and the predictions can be evaluated against each other for consistency and error-checking. The multi-level prediction framework achieves promising results on a novel surgery tool dataset and surgery knowledge base, which are important contributions of our work. This framework provides a viable solution for intelligent management of surgical tools in a hospital, potentially leading to significant cost savings and increased efficiencies.

Keywords: Surgical tool dataset · Multi-level predictions · Hierarchical classification · Surgery knowledge base

1 Introduction

Surgical tool and tray management is recognized as a difficult issue in hospitals worldwide. Stockert and Langerman [16] observed 49 surgical procedures involving over two-hundred surgery instrument trays, and discovered missing, incorrect or broken instruments in 40 trays, or in 20% of the sets. Guedon et al. [5] found equipment issues in 16% of surgical procedures; 40% was due to unavailability of a specific surgical tool when needed. Zhu et al. [24] estimated that 44% of packaging errors in surgical trays at a Chinese hospital were caused by packing the wrong instrument, even by experienced operators. This is significant given the volumes; for example, just one US medical institution processed over one-hundred-thousand surgical trays and 2.5 million instruments annually [16].

There are tens of thousands of different surgical tools, with new tools constantly being introduced. Each tool differs in shape, size and complexity – often in very minor, subtle, and difficult to discern ways, as shown in Fig. 1. Surgical sets, which can contain 200 surgical tools, are currently assembled manually [11]

© Springer Nature Switzerland AG 2021
M. Reyes et al. (Eds.): iMIMIC 2021/TDA4MedicalData 2021, LNCS 12929, pp. 3–12, 2021.
https://doi.org/10.1007/978-3-030-87444-5_1

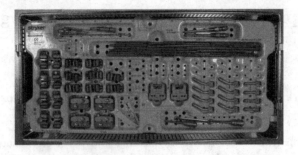

Fig. 1. Surgical tools - Hoffman Compact instruments and implants

but this is a difficult task even for experienced packing technicians. Given that surgical tool availability is a mission-critical task, vital to the smooth functioning of a surgery, ensuring that the tool is identified accurately is extremely important. Al Hajj et al. [2] reviewed convolutional neural network (CNN) architectures and a range of imaging modalities, applications, tasks, algorithms and detection pipelines used for surgical segmentation. They pointed out that hand crafted and hand engineered features had also been used for this task, and Bouget et al. [3] reviewed predominant features used for object-specific learning with surgical tools, and listed colour, texture, gradient and shape as being important for detection and classification. Yang et al. [19] presented a review of the literature regarding image-based laparoscopic tool detection and tracking using CNNs, including a discussion of available datasets and CNN-based detection and tracking methods. While CNNs can therefore provide viable solutions for surgical tool management, understanding how the CNN makes a prediction is important for building trust and confidence in the system.

Interpretability of predictions is then a critical issue – Rudin et al. [12] stated that interpretable machine learning is about models that are understood by humans, and interpretability can be achieved via separation of information as it traverses through the CNN models. Zhang et al. [21] developed an interpretable model that provided explicit knowledge representations in the convolutional layers (conv-layers) to explain the patterns that the model used for predictions. Linking middle-layer CNN features with semantic concepts for predictions provided interpretation for the CNN output [15,22,23]. How mid-level features of a CNN represent specific features of surgical tools and how they can provide hierarchical predictions is the focus of our work. CNNs learn different features of images at different layers, with higher layers extracting more discriminative features [20]. By associating feature maps at different CNN levels to levels in a hierarchical tree, a CNN model could incorporate knowledge of hierarchical categories for better classification accuracy. The model developed by Ferreira et al. [4] addressed predictions across five categorisation levels: gender, family, category, sub-category and attribute. The levels constituted a hierarchical structure, which was incorporated in the model for better predictions. The benefit of this hierarchical and interpretable approach for surgical tool management is that end users can then make rational, well reasoned decision on whether they can trust the information presented to them [12].

Wang et al. [18] discussed an approach to fine tuning that used wider or deeper layers of a network, and demonstrated that this significantly outperformed the traditional approaches which used pre-trained weights for fine-tuning. Going deeper was accomplished by constructing new top or adaptation layers, thereby permitting novel compositions without needing modifications to the pre-trained layers for a new task. Shermin et al. [14] showed that increasing network depth beyond pre-trained layers improved results for fine-grained and coarse classification tasks. We build on these approaches in our multi-level predictor.

Table 1. Surgical datasets

Characteristic	CATARACTS	Cholec80	Surgical tools
Size or Instances	50 videos	80 Videos	18300 images
Database Focus	Cataract Surgeries	Cholecystectomy Surgeries	Orthopaedics and General Surgery
Type of Surgery	Open Surgery	Laparoscopic	Open Surgery
Default Task	Detection	Detection	Classification
Type of Item	Videos	Videos	RGB Images
Number of Classes	21	7	361
Images Background	Tissue	Tissue	Flat colours
Image Acquisition Platform/Device	Toshiba 180I camera and MediCap USB200 recorder	Not Specified	Canon D-80 Camera and Logitech 922 Pro Stream Webcam
Image Illumination	Microscope Illumination	Fibre-optic in-cavity	Natural Light, LED, Fluorescent
Distance to Object	V.Close - Microscope	Close - in-cavity	30-cms to 60-cms
Annotations	Binary	Bounding Boxes	Multiple level
Dataset Organisation	500,000 frames each in Training and Test Sets	86,304 & 98,194 frames in Training and Test Set	14,640 images in Training and 3,660 in Validation set
Structure	Flat	Flat	Hierarchical
Image Resolution	1920 × 1080 pixels	Not Specified	600 × 400 pixels

2 Surgical Tool Dataset Overview

Kohli et al. [7] and Maier-Hein et al. [10] discussed the problems faced by the machine learning community stemming from a lack of data for medical image evaluation, which significantly impairs research in this area. There is just not enough high quality, well annotated data, representative of the particular surgery– a shortfall that needs to be addressed. Most medical datasets are one-off solutions for specific research projects, with limited coverage and restricted in numbers of images or data points [10]. To address this, we plan to create and curate a surgical tool dataset with tens of thousands of tool images across all surgical specialities with high quality annotations and reliable ground-truth information. Since surgery is organised along specialities, each with its own categories, a hierarchical classification of surgical tools would be extremely valuable. We therefore developed our initial surgical dataset with a hierarchical structure based on the surgical speciality, pack, set and tool. We captured RGB images of surgical tools using a DSLR camera and a webcam and tried to provide consideration to achieving viewpoint invariant object detection with different backgrounds, illumination, pose, occlusion and intra-class variations captured in the images. We focused on two specialities – Orthopaedics and General Surgery – of the 14 specialities reported by the American College of Surgeons [1]. The former offers a wide range of instruments and implants, while the latter covers the most common surgical tools. We propose to add the other specialities in a phased manner, and will make the dataset publicly available to facilitate research in this area.

CNNs have been successfully used for the detection, segmentation and recognition of surgical tools [9]. However, the datasets currently available for surgical tool detection present very small instrument sets; to illustrate this, the Cholec80, EndoVis 2017 and m2cai16-tool datasets have seven instruments, the CATARACTS dataset has 21 instruments, the NeuroID dataset has eight instruments and the LapGyn4 Tool Dataset has three instruments [2,17]. While designing CNNs to recognise seven or eight instruments for research purposes may be justifiable, this is nowhere nearly adequate enough for real work conditions. Any model trained using this data is unlikely to be usable anywhere else, not even in the same hospital six months later. We needed to develop a new dataset for our work as these surgical tool datasets did not offer a sufficiently large variety or number of tools for analysis, nor were they arranged hierarchically. A comparison of our dataset with CATARACTS [2] and Cholec80 [17], two important publicly available datasets, is presented in Table 1.

Table 2. Surgery knowledge base (excerpt)

Speciality	Pack	Set	Tool
Orthopaedics	VA Clavicle Plating Set	LCP Clavicle Plates	Clavicle Plate 3.5 8 Hole Right
Orthopaedics	Trimed Wrist Fixation System	Fixation Fragment Specific	Dorsal Buttress Pin 26 mm
General Surgery	Cutting & Dissecting	Scissors	9 Metzenbaum Scissors
General Surgery	Clamping & Occluding	Forceps	6 Babcock Tissue Forceps

2.1 Surgery Knowledge Base

Setti [13] points out that most public benchmark datasets only provide images and label annotations, but providing additional prior knowledge can boost performance of CNNs. To complement the dataset, we developed a comprehensive surgery knowledge-base (Table 2) as an attribute-matrix which makes rich information available to the training regime. This proved to be a convenient and useful data structure that captures rich information of class attributes – or the nameable properties of classes – and makes it readily available for computational reasoning [8]. We developed the knowledge representation structure for 18,300 images to provide rich, multi-level and comprehensive information about each image. The attribute matrix data structure proved to be easy to work with, simple to change and update, and it also provided computational efficiencies.

3 Experimental Method

We implemented our project in Tensorflow v-2.4.1 and Keras v-2.4.3. Our architecture consists of a ResNet50V2 network [6] which we trained on the Surgical Tool training dataset, by replacing the top layer with a dropout and dense layer with 361 outputs. We initially did not use the knowledge base annotations, only the tool labels and trained with the configuration in Table 4 with early stopping on validation categorical accuracy. We were able to obtain good predictions from this model with accuracy score at 93.51%, but only at the tool level. We then used this pre-trained architecture with surgical tool weights as our base model, froze the base model, and added separate classification pipelines, one for each prediction of interest - speciality, set, pack and tool (See Fig. 2). We relied on the knowledge base annotations which provided data for two specialities, twelve packs, thirty-five sets and 361 possible tools, and used it to create data-frames for the training and validation data. Each image was associated with the relevant annotations for each output, in the form of columns of text values or categorical variables representing the multiple classes for each output. This multi-task framework effectively shared knowledge of the different attribute categories for each image or visual representation. We developed a custom data handler for the training data (x_set) and for the labels for each of the four outputs (y_cat, y_pack, y_set, y_tool), and used one hot encoding to represent the categorical variables in our model. We then implemented training and validation data generators based on our custom data handler to provide batches of data to the model. Our model was compiled with one input (image) and four outputs.

Table 3. Results - Val accuracy with output at different layers

All outputs at:	Total parameters	Parameters trained	Speciality	Pack	Set	Tool
Conv2_block1_1_relu	700,570	686,490	0.956	0.356	0.258	0.091
Conv3_block1_1_relu	1,210,266	948,634	0.989	0.621	0.507	0.231
Conv4_block1_1_relu	3,060,634	1,472,922	0.997	0.927	0.851	0.663
Conv5_block1_1_relu	11,625,370	2,521,498	0.999	0.975	0.945	0.890

We tested outputs at different layers to evaluate the impact of changing the depth of the network, with the results in Table 3. In each experiment, parameters

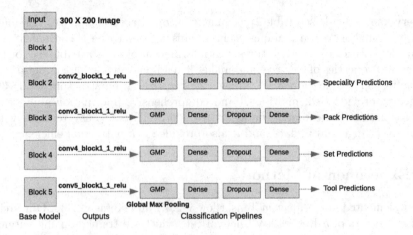

Fig. 2. Resnet50V2 architecture with multiple outputs

available and actually trained were controlled by adjusting the numbers of layers. An operation within a block in ResNet50V2 consisted of applying convolution, batch normalisation and activation to an input; we obtained our outputs after the first operation in each block. These outputs were fed to external global max pooling and dense layers. A dropout layer regulated training – we replaced this with a batch normalisation layer but results did not improve. Since this was a multi-class problem, a dense layer with softmax activation was used for the final classification of each prediction, customised to the relevant number of classes. As we expected, better results were obtained by including more layers and by training more parameters – best results were obtained by including all layers up to Block 5. However, it is noteworthy that high accuracy was obtained for specific predictions even early in the model – for example, predictions for speciality were at 95.6% by block 2, for pack and set were at 92.7% and 85.10% at block 4 and for tool at 89% at block 5. Clearly it was possible to obtain accurate predictions for higher level categories using early layers of the model. This is explored further with the objective of improving interpretability for the end user, while reducing the total number of parameters that needed to be trained in the model.

Table 4. Training configuration

Parameter	Optimiser	Learning rate	Batch size	Activation	Loss	Metric
Value	Adam	0.001	64	Softmax	Categorical crossentropy	Categorical accuracy

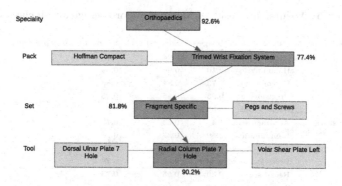

Fig. 3. Interpretable multi-level predictions

The training set images from the surgery dataset and annotations from the knowledge base were used for training, with real time training data augmentation – including horizontal flip, random contrast and random brightness operations. We used the configuration in Table 4, the initial learning rate of 0.001 was decreased to 0.0001 at epoch 45 and to 0.00005 at epoch 75. A dropout rate of 0.2 was imposed. We implemented early stopping on val loss with a patience of 20 epochs. The total parameters in the model were 10,511,258, and parameters trained were 1,407,386 in each of the experiments.

1. ImageNet Training: For an initial baseline experiment, we used a ResNet50V2 model with ImageNet weights and four separate classification outputs were trained, one for each hierarchy – speciality, set, pack and tool.
2. Surgical Tool Training: We used the pre-trained base model with surgical tool weights, and trained the model with its four classification pipelines using the configuration as in Table 4 and architecture as in Fig. 2.
3. Depth Adjusted Surgical Tool Training: We used the pre-trained model with surgical tool weights as before, but changed the levels within the blocks of the ResNet-50V2 model from which we obtained outputs, thereby adjusting the depth of training. The outputs from Block 5 and 2 were obtained from conv"x"_block1_1, and from Block 3 and 4 were from conv"x"_block4_2. We did this to evaluate the effects of changing depths on the prediction accuracy; this was a minor change within the block but the total number of parameters trained were maintained the same.

Table 5. Architecture results - Macro score or average for all classes

Level	Metric	ImageNet	Surgical-tools	Surgical-tools depth adjusted
Speciality score	Accuracy score	0.90	0.94	0.94
	Hamming loss	0.10	0.06	0.06
	f1 score	0.73	0.84	0.83
	Precision score	0.93	0.95	0.95
	Recall score	0.96	0.99	0.99
Pack score	Accuracy score	0.41	0.63	0.77
	Hamming loss	0.59	0.37	0.23
	f1 score	0.25	0.53	0.73
	Precision score	0.43	0.67	0.76
	Recall score	0.30	0.55	0.73
Set score	Accuracy score	0.31	0.84	0.89
	Hamming loss	0.69	0.16	0.11
	f1 score	0.24	0.79	0.84
	Precision score	0.36	0.82	0.85
	Recall score	0.25	0.80	0.87
Tool score	Accuracy score	0.20	0.90	0.90
	Hamming loss	0.80	0.10	0.10
	f1 score	0.16	0.86	0.86
	Precision score	0.78	0.91	0.91
	Recall score	0.27	0.91	0.90

4   Results and Conclusions

Our results, on a separate test subset of data, are shown in Table 5. The test data was images that the model had not seen before, as a sample of 400 random images across all classes had been reserved for testing. Training with ImageNet weights did not provide good results, but the use of surgical tool weights demonstrated that the model had captured relevant information about the dataset and was able to provide good predictions at multiple levels. In this architecture, by extracting multiple predictions along layers from coarse to fine as data traverses the CNN, early layers provided predictions corresponding to specialities while later layers provide finer predictions, such as tool classifications (Fig. 3). It was easy for the CNN to distinguish between our two speciality classes, since General Surgery tools are visually different from orthopaedic tools – as we add more specialities where the visual distinction is not so clear, we may need to train at deeper levels. As the classes increased to 12, 35 and 361 for pack, set and tool respectively, predictions from deeper layers were needed. These hierarchical predictions are expected to provide better interpretability since multiple predictions can be tested and evaluated against each other for consistency or error by the end user. Adjusting the depths of layers used as outputs for predictions improved the results, even within the same block, demonstrating that more features are learned as the data travels through the CNN layers.

We developed a CNN framework that successfully utilised the hierarchical nature of surgical tool classes to provide a comprehensive set of classifications for each tool. This framework was deployed and tested on a new surgical tool dataset and knowledge base. The multi-level prediction system provides a good solution for classification of other types of medical images, if they are hierarchically organised with a large number of classes.

References

1. ACS: What are the surgical specialties? (2021). https://www.facs.org/education/resources/medical-students/faq/specialties. Accessed 15 Feb 2021
2. Al Hajj, H., et al.: CATARACTS: challenge on automatic tool annotation for cataRACT surgery. Med. Image Anal. **52**, 24–41 (Feb 2019). https://doi.org/10.1016/j.media.2018.11.008
3. Bouget, D., Allan, M., Stoyanov, D., Jannin, P.: Vision-based and marker-less surgical tool detection and tracking: a review of the literature. Med. Image Anal. **35**, 633–654 (Jan 2017). https://doi.org/10.1016/j.media.2016.09.003
4. Ferreira, B., Baia, L., Faria, J., Sousa, R.: A unified model with structured output for fashion images classification. In: AI for Fashion - The Third International Workshop on Fashion and KDD, London, United Kingdom (2018)
5. Guedon, A.C., et al.: Where are my instruments? Hazards in delivery of surgical instruments. Surg. Endosc. **30**(7), 2728–2735 (2016). https://doi.org/10.1007/s00464-015-4537-7
6. He, K., Zhang, X., Ren, S., Sun, J.: Deep residual learning for image recognition. In: IEEE Conference on Computer Vision and Pattern Recognition (CVPR), pp. 770–778. Washington (DC). IEEE Computer Society (2016). https://doi.org/10.1109/CVPR.2016.90
7. Kohli, M.D., Summers, R.M., Geis, J.R.: Medical image data and datasets in the era of machine learning - white paper from the 2016 C-MIMI meeting dataset session. J. Digit. Imaging **30**, 392–399 (2017). https://doi.org/10.1007/s10278-017-9976-3
8. Lampert, C.H., Nickisch, H., Harmeling, S.: Attribute-based classification for zero-shot visual object categorization. IEEE Trans. Pattern Anal. Mach. Intell. **36**(3), 453–465 (March 2014). https://doi.org/10.1109/TPAMI.2013.140
9. LeCun, Y., Bengio, Y., Hinton, G.: Deep learning. Nature **521**(1038), 436–444 (2015). https://doi.org/10.1038/nature14539
10. Maier-Hein, L., Eisenmann, M., Sarikaya, D., Marz, K., et al.: Surgical data science - from concepts to clinical translation. ArXiv, abs/2011.02284 (2020)
11. Mhlaba, J.M., Stockert, E.W., Coronel, M., Langerman, A.J.: Surgical instrumentation: the true cost of instrument trays and a potential strategy for optimization. J. Hosp. Admin. **4**(6), 82–88 (2015). https://doi.org/10.5430/jha.v4n6p82
12. Rudin, C., Chen, C., Chen, Z., Huang, H., Semenova, L., Zhong, C.: Interpretable machine learning: fundamental principles and 10 grand challenges. ArXiv, abs/2103.11251 (2021)
13. Setti, F.: To know and to learn - about the integration of knowledge representation and deep learning for fine-grained visual categorization. In: 13th International Joint Conference on Computer Vision, Imaging and Computer Graphics Theory and Applications (VISIGRAPP). vol 5. pp. 387–392 (2018). https://doi.org/10.5220/0006651803870392
14. Shermin, T., Murshed, M., Teng, S., Lu, G.: Depth augmented networks with optimal fine-tuning. ArXiv, abs/1903.10150 (2019)
15. Simon, M., Rodner, E.: Neural activation constellations: unsupervised part model discovery with convolutional networks. In: International Conference on Computer Vision (ICCV). 1143–1151 (2015) https://doi.org/10.1109/ICCV.2015.136

16. Stockert, E.W., Langerman, A.J.: Assessing the magnitude and costs of intraoperative inefficiencies attributable to surgical instrument trays. J. Am. Coll. Surg. **219**(4), 646–655 (Oct 2014). https://doi.org/10.1016/j.jamcollsurg

17. Twinanda, A.P., Shehata, S., Mutter, D., Marescaux, J., de Mathelin, M., Padoy, N.: EndoNet: a deep architecture for recognition tasks on laparoscopic videos. In: IEEE Trans. Med. Imaging **36**(1), 86–97 (Jan 2017). https://doi.org/10.1109/TMI. 2016.2593957

18. Wang, Y.-X., Ramanan, D., Hebert, M.: Growing a brain: fine-tuning by increasing model capacity. In: IEEE Conference on Computer Vision and Pattern Recognition (CVPR), 3029–3038 (2017)

19. Yang, C., Zhao, Z., Hu, S.: Image-based laparoscopic tool detection and tracking using convolutional neural networks: a review of the literature. Comput. Assist. Surg. **25**(1), 15–28 (Dec 2020). https://doi.org/10.1080/24699322.2020.1801842

20. Zeiler, M.D., Fergus, R.: Visualizing and understanding convolutional networks. In: Fleet, D., Pajdla, T., Schiele, B., Tuytelaars, T. (eds.) ECCV 2014. LNCS, vol. 8689, pp. 818–833. Springer, Cham (2014). https://doi.org/10.1007/978-3-319-10590-1_53

21. Zhang, Q., Wang, X., Wu, Y.N., Zhou, H., Zhu, S.C.: Interpretable CNNs for object classification. In: IEEE Trans. Pattern Anal. Mach. Intell. **43**(10) 3416–3431 (1 Oct 2020). https://doi.org/10.1109/TPAMI.2020.2982882

22. Zhang, Q., Yang, Y., Ma, H., Wu, Y.N.: Interpreting CNNs via decision trees. In: IEEE/CVF Conference on Computer Vision and Pattern Recognition (CVPR), pp. 6254-6263 (2019). Long Beach, CA, USA, https://doi.org/10.1109/CVPR.2019. 00642

23. Zhou, B., Khosla, A., Lapedriza, A., Oliva, A., Torralba, A.: Object detectors emerge in deep scene CNNs. In: International Conference on Learning Representations, ICLR 2015, San Diego, CA, USA (2015)

24. Zhu, X., Yuan, L., Li, T., Cheng, P.: Errors in packaging surgical instruments based on a surgical instrument tracking system: an observational study. BMC Health Serv. Res. **19**(1), 176 (2019). https://doi.org/10.1186/s12913-019-4007-3

Soft Attention Improves Skin Cancer Classification Performance

Soumyya Kanti Datta[✉] ⓘ, Mohammad Abuzar Shaikhⓘ, Sargur N. Srihari, and Mingchen Gaoⓘ

State University of New York at Buffalo, Buffalo, USA
{soumyyak,mshaikh2,srihari,mgao8}@buffalo.edu

Abstract. In clinical applications, neural networks must focus on and highlight the most important parts of an input image. Soft-Attention mechanism enables a neural network to achieve this goal. This paper investigates the effectiveness of Soft-Attention in deep neural architectures. The central aim of Soft-Attention is to boost the value of important features and suppress the noise-inducing features. We compare the performance of VGG, ResNet, Inception ResNet v2 and DenseNet architectures with and without the Soft-Attention mechanism, while classifying skin lesions. The original network when coupled with Soft-Attention outperforms the baseline [16] by 4.7% while achieving a precision of 93.7% on HAM10000 dataset [25]. Additionally, Soft-Attention coupling improves the sensitivity score by 3.8% compared to baseline [31] and achieves 91.6% on ISIC-2017 dataset [2]. The code is publicly available at github (https://github.com/skrantidatta/Attention-based-Skin-Cancer-Classification).

1 Introduction

Skin cancer is the most common cancer and one of the leading causes of death worldwide. Every day, more than 9500 people[1] in the United States are diagnosed with skin cancer, with 3.6 million people[2] diagnosed with basal cell skin cancer each year. Early diagnosis of the illness has a significant effect on the patients' survival rates. As a result, detecting and classifying skin cancer is important.

It is difficult to distinguish between malignant and benign skin diseases because they look so similar. Although a dermatologist's visual examination is the first step in detecting and diagnosing a suspicious skin lesion, it is usually followed by dermoscopy imaging for further analysis [32]. Dermoscopy images

---

S. K. Datta and M. A. Shaikh are equal contributors.

[1] https://www.skincancer.org/skin-cancer-information/skin-cancer-facts/.
[2] https://www.skincancer.org/skin-cancer-information/basal-cell-carcinoma/.

Electronic supplementary material The online version of this chapter (https://doi.org/10.1007/978-3-030-87444-5_2) contains supplementary material, which is available to authorized users.

M. Reyes et al. (Eds.): iMIMIC 2021/TDA4MedicalData 2021, LNCS 12929, pp. 13–23, 2021.
https://doi.org/10.1007/978-3-030-87444-5_2

provide a high-resolution magnified image of the infected skin region, but they are not without their drawbacks. Due to the image size being large, it becomes difficult for the feature extractors to extract out the relevant features for classification. Various methods such as Segmentation and detection, Transfer learning, General Adversarial networks, etc. have been used to detect and classify skin cancer. Despite significant progress, skin cancer classification is still a difficult task. This is due to the lack of annotated data and low inter-class variation. Furthermore, the task is complicated by contrast variations, color, shape, and size of the skin lesion, as well as the presence of various artifacts such as hair and veins. Inspired by the work done in [18], this paper studies the effect of soft attention mechanism in deep neural networks. Deep learning architectures identify the image class by learning the salient features and nonlinear interactions. The soft-attention mechanism improves performance by focusing primarily on relevant areas of the input. Moreover, the soft-attention mechanism makes the image classification process transparent to medical personnel, as it maps the parts of the input that the network uses to classify the image, thereby, increasing trust in the classification model.

Following Krichevsky [12], large-scale image classification tasks using deep convolutional neural networks have become common. As reported in the paper [3], the task of skin cancer classification using images has improved rapidly since the implementation of Deep Neural Networks. To make progress, we suggest that soft attention be used to identify fine-grained variability in the visual features of skin lesions.

Existing art in the field of skin cancer classification used streamlined pipelines based upon current Computer Vision [4]. Masood et al. in their paper [13] proposed a general framework from the viewpoint of computer vision, where the methods such as calibration, preprocessing, segmentation, balancing of classes and cross validation are used for automated melanoma screening. In 2018, Valle et al. [26] investigated ten different methodologies to evaluate deep learning models for skin lesion classification. Data augmentation, model architecture, image resolution, input normalization, train dataset, use of segmentation, test data augmentation, additional use of support vector machines, and use of transfer learning are among the ten methodologies they evaluated. They stated that data augmentation had the greatest impact on model efficiency. The same observation is confirmed by Perez's 2018 paper "Data Augmentation for Skin Lesion Analysis" [15].

Nonetheless, the problems of low inter-class variance and class imbalance in skin lesion image datasets remain, seriously limiting the capabilities of deep learning models [30]. To fix the lack of annotated data, Zunair et al. [32] proposed the use of adversarial training and Bissoto et al. [1] proposed the use of Generative Adversarial Networks to produce realistic synthetic skin lesion photos. Zhang et al. [31] in 2019 proposed the attention residual learning convolutional neural network for skin lesion classification which is based on self attention mechanism.

In this paper, we suggest using a Soft Attention mechanism in conjunction with a Deep Convolutional Neural Network (DCNN) to classify skin cancer. Rather than using attention modules with residual blocks and stacking them one after another like in paper [31], we integrated the soft attention module into the various DCNN architectures such as Inception ResNet v2 [22], which improved the performance of those architectures. Our model used the DCNN to extract the features maps from the skin lesion images, and the soft attention module assisted the DCNN in focusing more on the important features of the images without completely discarding the other features. Our paper's primary contribution is that we offer a unique technique for integrating soft attention mechanism with DCNNs to optimize performance, and we outperformed the state-of-the-art on skin lesion classification on the HAM10000 dataset [25] and ISIC-2017 dataset [2].

2   Method

In this paper, five deep neural networks which are ResNet34, ResNet50 [6], Inception ResNet v2 [22], DenseNet201 [8] and VGG16 [20], are implemented with soft attention mechanism, to classify skin cancer images. ResNet34, ResNet50 [6], Inception ResNet v2, DenseNet201 [8] and VGG16 [20] are all state of the art feature extractors which are trained on ImageNet dataset. The main components and architecture of the proposed approach is described below:

2.1   Dataset

The experiment is performed on two datasets separately. The two datasets are as follows: HAM10000 dataset [25] and ISIC 2017 dataset [2].

The HAM10000 dataset [25] consists of 10015 dermatoscopic images of a size of 450 × 600. It consists of 7 diagnostic categories as follows: Melanoma(MEL), Melanocytic Nevi (NV), Basal Cell Carcinoma (BCC), Actinic Keratosis, and Intra-Epithelial Carcinoma (AKIEC), Benign Keratosis (BKL), Dermatofibroma (DF), Vascular lesions (VASC). All the images are resized to 299 × 299 for Inception ResNet v2 [22] architecture and 224 × 224 for the other architectures.

The ISIC 2017 dataset consists of 2600 images. In the training dataset there are 2000 images of 3 catagories as follows: benign nevi, seborrheic keratosis, and melanoma. The test dataset consist of 600 images. In this experiment we are training our model to classify only benign nevi and seborrheic keratosis. All the images resized to 224 × 224.

The data in both datasets is then cleaned to remove class imbalances. This is done by the process of over-sampling and under-sampling of data so that there are equal number of images per class. The images are then normalized by dividing each pixel with 255 to keep the pixel values in the range 0 to 1.

2.2   Soft Attention

When it comes to skin lesion images, only a small percentage of pixels are relevant as the rest of the image is filled with various irrelevant artifacts such as veins and

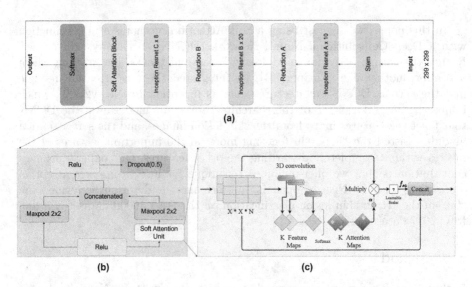

Fig. 1. (a) End to end architecture of Inception ResNet v2 [22] with Soft Attention block. **(b)** The schema for Soft Attention block. **(c)** Soft Attention unit

hair. So, to focus more on these relevant features of the image, soft attention is implemented. Inspired by the work proposed by Xu et al. [28], for image caption generation and the work done by Shaikh et al. [18], where they used attention mechanism on images for handwriting verification, in this paper, soft attention is used to classify skin cancer.

In Fig. 2, we can see that areas with higher attention are red in color. This is because soft attention discredits irrelevant areas of the image by multiplying the corresponding feature maps with low weights. Thus the low attention areas have weights closer to 0. With more focused information, the model performs better.

In the soft attention module as discussed in paper [18] and [23], the feature tensor (t) which flows down the deep neural network is used as input.

$$f_{sa} = \gamma t((\sum_{k=1}^{K} softmax(W_k * t))) \tag{1}$$

This feature tensor $t \in \mathbb{R}^{h \times w \times d}$ is input to a 3D convolution layer [24] with weights $W_k \in \mathbb{R}^{h \times w \times d \times K}$, where K is the number of 3D weights. The output of this convolution is normalized using softmax function to generate $K = 16$ attention maps. As shown in Fig. 1(c), these attention maps are aggregated to produce a unified attention map that acts as a weighting function α. This α is then multiplied with t to attentively scale the salient feature values, which is further scaled by γ a learnable scalar. Because various images require different γ values, γ is treated as a learnable parameter and not hard-coded. This allows the network to determine how much it should focus on the attention maps on its own. Finally, the attentively scaled features (f_{sa}) are concatenated with the original feature t in form of a residual branch. During training we initialize γ from 0.01 so that the network can slowly learn to regulate the amount of attention required by the network.

2.3 Model Setup

In this section, the detailed architecture of the models is discussed. For all experiments, to train the networks, Adam optimizer [11] of 0.01 learning rate and 0.1 epsilon is used. A batch normalization [10] layer is added after each layer in all the networks to introduce some regularization. For the HAM10000 dataset [25], since there are 7 classes of skin cancer, an output layer with 7 hidden units is implemented, followed by a softmax activation unit. We employed a batch size of 16 during both training and testing. In this paper, the model is evaluated using $Precision = \frac{TP}{TP+FP}$, $Accuracy = \frac{TP+TN}{T}$, $Sensitivity = \frac{TP}{TP+FN}$, $Specificity = \frac{TN}{TN+FP}$ and AUC scores [9]. Here TN, TP, FP, FN, T mean, True Negatives, True Positives, False Positives, False Negatives, Total Number respectively. All the experiments were executed on the Keras framework with tensorflow version 2.4.0.

Inception ResNet V2: In Inception ResNet v2 [22], the soft attention layer is added to the Inception Resnet C block of the model where the feature size of the image is 8×8 as shown in Fig. 1(a). In this case, the soft attention layer is followed by a maxpool layer with a pool size of 2×2, which is then concatenated with the filter concatenate layer of the inception block. The concatenate layer is then followed by a relu activation unit. To regularize the output of the attention layer, the activation unit is followed by a dropout layer [21] with dropout probability of 0.5 as shown in Fig. 1(b). The network is trained for 150 epochs with early stopping patience of 30 epochs while monitoring the validation loss for a minimum delta of 0.001. The dropout and early stopping regularization prevents the network from over-fitting to the training dataset. The overall network is shown in Fig. 1(a). The other architectures, such as ResNet34, ResNet50 [6], DenseNet201 [8], and VGG16 [20], and the process by which the Soft Attention block was integrated with them, are described in the supplementary document.

2.4 Loss Function

In this experiment, there are seven different classes of skin cancer. Hence, categorical cross entropy loss (L_{CCE}) is used to optimize the neural network.

$$L_{CCE} = -\sum_{i=1}^{C} t_i log(f(s)_i) \tag{2}$$

where

$$f(s)_i = \frac{e^{s_i}}{\sum_{j=1}^{C} e^{s_j}} \tag{3}$$

Here, as there are seven classes, $C \in [0..6]$, where t_i is the ground truth and s_i is the CNN score for each class i in C. $f(s)_i$ is the softmax activation function applied to the scores.

Table 1. Ablation results for choosing the best model on HAM10000 dataset [25]. [22] refers to IRv2 architecture, [8] refers to DenseNet 201 architecture, [20] refers to VGG 16 architecture, and [6] refers to ResNet architecture.

Dis.	Precision										AUC										#
	[22]	[22]+SA	[8]	[8]+SA	[20]	[20]+SA	[6]50	[6]50+SA	[6]34	[6]34+SA	[22]	[22]+SA	[8]	[8]+SA	[20]	[20]+SA	[6]50	[6]50+SA	[6]34	[6]34+SA	
AKIEC	0.830	1.000	1.000	0.920	0.620	0.700	0.740	0.670	0.670	0.500	0.993	0.981	0.975	0.967	0.949	0.964	0.980	0.981	0.969	0.970	23
BCC	0.850	0.880	0.830	0.800	0.540	0.620	0.910	0.880	0.660	0.880	0.997	0.998	0.993	0.994	0.977	0.984	0.997	0.996	0.991	0.993	26
BKL	0.850	0.720	0.690	0.730	0.570	0.630	0.670	0.670	0.510	0.520	0.970	0.982	0.960	0.964	0.930	0.900	0.948	0.964	0.904	0.916	66
DF	0.670	1.000	0.500	1.000	0.250	0.500	0.800	1.000	0.400	0.330	0.973	0.982	0.851	0.921	0.847	0.809	0.973	0.971	0.925	0.949	6
MEL	0.700	0.670	0.540	0.530	0.500	0.430	0.520	0.730	0.420	0.540	0.965	0.974	0.963	0.976	0.925	0.956	0.961	0.973	0.910	0.953	34
NV	0.930	0.970	0.950	0.950	0.930	0.950	0.950	0.950	0.930	0.930	0.984	0.984	0.975	0.976	0.954	0.951	0.974	0.979	0.944	0.958	663
VASC	1.000	1.000	0.900	0.830	1.000	1.000	0.900	1.000	0.910	0.820	1.000	1.000	0.993	0.999	0.972	0.999	0.995	0.999	0.999	0.996	10
Avg	0.832	0.892	0.771	0.824	0.631	0.690	0.783	0.841	0.642	0.646	0.983	0.984	0.959	0.971	0.936	0.937	0.975	0.980	0.949	0.962	828
W. Avg	0.905	0.937	0.904	0.909	0.862	0.882	0.898	0.910	0.857	0.865	0.982	0.984	0.974	0.975	0.951	0.948	0.972	0.978	0.942	0.957	828

3   Results

3.1   Ablation Analysis

Table 1 lists, the performance of all the models in terms of precision, and AUC score on HAM10000 dataset [25]. In this table (+SA) stands for models with soft attention. From the table, it can be observed that IRv2 when coupled with SA (IRv2 [22] + SA) shows significant improvements in results, with a precision and AUC score of 93.7% and 98.4% respectively, which are also the highest scores amongst all models. Furthermore, we can see that Soft Attention (SA) boosts the performance of IRv2 by 3.2% in terms of precision as compared to the original IRv2 model. This phenomenon is true for VGG16, ResNet34, ResNet50 and DenseNet201 as well. For instance, Soft Attention (SA) boosts the precision of DenseNet201 [8], ResNet34 [6], ResNet50 [6], and VGG16 [20] by 0.5%, 0.8%, 1.2% and 2% respectively. We see a similar behaviour for the AUC scores when SA block is integrated in to the networks, such as, the performance of ResNet50 [6], and ResNet34 [6] has grown by 0.6% and 1.5% respectively and the performance of DenseNet201 [8], and VGG16 [20] is on par with the original models.

Although IRv2+SA performs the best in terms of weighted average (W.Avg), when we look at it's class wise performance, we can see that Soft Attention enhances the efficiency of the original IRv2 while categorizing AKIEC, BCC, DF and NV by 17%, 3%, 33% and 4% respectively in terms of precision. Moreover, when comparing AUC scores, the IRv2+SA performs better for BKL and

MEL by 1.2% and 0.9% respectively, while, for BCC, NV and VASC, IRv2+SA performs as good as original model.

We thus select IRv2 coupled with SA (IRv2+SA) for our experiments, also the SA block consistently boosts the performance of it's original counterpart, hence, we can justify the integration of Soft Attention to the networks.

3.2 Quantitative Analysis

The proposed approach is compared with state-of-the-art models for skin cancer classification on the HAM10000 dataset [25] in Table 2(a). Our Soft Attention-based approach outperforms the baseline [16] by 4.7% in terms of precision. In terms of AUC scores, our Soft Attention-based approach clearly outperforms them all by 0.5% to 4.3%.

Table 2. (a) Comparison with state-of-the-art-model in terms of average AUC score, precision and accuracy on HAM10000 dataset [25]. **(b)** Comparison with state-of-the-art-model in terms of AUC, accuracy, sensitivity and specificity score on ISIC-2017 dataset [2] for seborrheic keratosis classification

Model	Avg AUC	Precision	Accuracy
Loss balancing and ensemble [5]	0.941	–	0.926
Single model deep learning [29]	0.974	–	0.864
Data classification augmentation [19]	0.975	–	0.853
Two path CNN model [14]	–	–	0.886
Various deep CNN (baseline) [16]	0.979	0.890	–
IRv2+SA(Proposed Approach)	0.984	0.937	0.934

(a)

Networks	AUC	Accuracy	Sensitivity	Specificity
ResNet50 [6]	0.948	0.842	0.867	0.837
RAN50 [27]	0.942	0.862	0.878	0.859
SEnet50 [7]	0.952	0.863	0.856	0.865
ARL-CNN50 [31]	0.958	0.868	0.878	**0.867**
(IRv2$_{12x12}$+SA)vs. N&M	0.922	0.890	**0.956**	0.589
(IRv2$_{5x5}$+SA)vs. N&M	0.942	0.900	0.932	0.687
(IRv2$_{12x12}$+SA)vs. N	0.935	0.898	0.945	0.711
(IRv2$_{5x5}$+SA)vs. N	0.959	0.904	0.916	0.833

(b)

We also tested the model with different train-test splits on the HAM10000 dataset [25], we discovered that the model with 85% training data outperforms the model with 80% and 70% training data by 2.2% and 2.6% respectively, as shown in supplementary material's Table 1. Hence we select 85/15% training/testing split for performing our experiments. In Table 2(b), the performance of the proposed approach Inception Resnet V2 [22] (IRv2$_{5x5}$+SA and IRv2$_{12x12}$+SA) with soft attention is measured on ISIC-2017 dataset [2] on basis of AUC scores, Accuracy, Sensitivity and Specificity with the state-of-the-art models. Here (IRv2$_{5x5}$+SA vs. N and IRv2$_{12x12}$+SA vs. N) refers to classification of seborrheic keratosis with respect to only benign nevi, and (IRv2$_{5x5}$+SA vs. N&M and IRv2$_{12x12}$+SA vs. N&M) refers to classification of seborrheic keratosis with respect to both benign nevi and melanoma.

From Table 2(b), it can be observed that in IRv2$_{5x5}$+SA, and in IRv2$_{12x12}$+SA, the attention layer was added when the feature map size is 5×5 and 12×12 respectively. Out of the four models with soft attention, the model $IRv2_{5\times5}+SA$ vs. N outperforms $IRv2_{12\times12}+SA$ vs. N, $IRv2_{5\times5}+SA$ vs. N&M and $IRv2_{12\times12}+SA$ vs. N&M, in terms of AUC scores by 1.7% to 3.7%, Accuracy by 0.4% to 1.4%, and Specificity by a percentage of 12.2% to 24.4% respectively whereas $IRv2_{12\times12}+SA$ vs. N&M outperforms $IRv2_{5\times5}+SA$ vs. N, $IRv2_{5\times5}+SA$ vs. N&M and $IRv2_{12\times12}+SA$ vs. N in terms of Sensitivity by 4.0%, 2.4% and 1.1% respectively. When $IRv2_{5\times5}+SA$ vs. N is compared with the ARL-CNN50 [31] (baseline model), it performs on par with it in terms of AUC score but our model outperforms it when it comes to accuracy and Sensitivity by 3.6% and 3.8% respectively. But ARL-CNN50 [31] takes the upper hand when it comes to Specificity by 3.4%. Since sensitivity measures the proportion of correctly identified positives and specificity measures the proportion of correctly identified negatives, we are prioritizing Sensitivity because classifying a person with cancer as not having cancer is riskier than vice versa.

3.3   Qualitative Analysis

Figure 2 displays the Soft Attention heat maps from the IRv2+SA model. In the Fig. 2, the images on the bottom row are the input images from the HAM10000 dataset [25]. The images in the middle row under the columns SA Map show the Soft Attention maps superimposed on input images to show where the model is focusing and the images of the top row are attention maps themselves.

In Fig. 2, we show pairs of comparison between the Soft Attention maps with Grad-CAM [17] heatmaps. In the first pair, the SA map focuses on the main part of the lesion area whereas the Grad-cam heatmap is slightly shifted towards top left and is also spread out on the uninfected area of skin. We have similar observations for the second and third pairs as well. From this observation it is

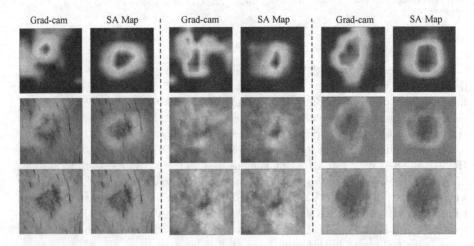

Fig. 2. Comparison of GradCAM [17] heatmaps with our Soft Attention (SA) maps on HAM10000 dataset [25]

evident that the Soft Attention maps are focused more on the relevant locations of the image compared to Grad-CAM [17] heatmaps.

4   Conclusion

In this paper, we present the implementation and utility of Soft Attention mechanism being applied while image encoding to tackle the problem of high-resolution skin cancer image classification. The model outperformed the current state-of-the-art approaches on the HAM10000 dataset [25] and the ISIC-2017 dataset [2]. This demonstrates the Soft Attention based deep learning architecture's potential and effectiveness in image analysis as well as in skin cancer classification. The Soft Attention mechanism also eliminates the need of using external mechanisms like GradCAM [17], and internally provides the location of where the model focuses while categorizing a disease, while also boosting the performance of the main network. Soft Attention has the added advantage of naturally dealing with image noise internally.

In future, we believe the salient regions proposed by Soft Attention can be used as salient regions for downstream tasks like classification, Visual Q&A and captioning as it will benefit datasets that don't have any bounding box annotations. Furthermore, this model can be also implemented in dermoscopy systems to assist dermatologists. Lastly, this mechanism can easily be implemented to classify data from other medical databases as well.

References

1. Bissoto, A., Perez, F., Valle, E., Avila, S.: Skin lesion synthesis with generative adversarial networks. In: Stoyanov, D., et al. (eds.) CARE/CLIP/OR 2.0/ISIC - 2018. LNCS, vol. 11041, pp. 294–302. Springer, Cham (2018). https://doi.org/10.1007/978-3-030-01201-4_32
2. Codella, N.C.F., et al.: Skin lesion analysis toward melanoma detection: a challenge at the 2017 international symposium on biomedical imaging (ISBI), hosted by the international skin imaging collaboration (ISIC). CoRR abs/1710.05006 (2017). http://arxiv.org/abs/1710.05006
3. Esteva, A., et al.: Dermatologist-level classification of skin cancer with deep neural networks. Nature **542**(7639), 115–118 (2017)
4. Fornaciali, M., Carvalho, M., Bittencourt, F.V., Avila, S., Valle, E.: Towards automated melanoma screening: proper computer vision & reliable results. arXiv preprint arXiv:1604.04024 (2016)
5. Gessert, N., Nielsen, M., Shaikh, M., Werner, R., Schlaefer, A.: Skin lesion classification using ensembles of multi-resolution EfficientNets with meta data. MethodsX **7**, 100864 (2020)
6. He, K., Zhang, X., Ren, S., Sun, J.: Deep residual learning for image recognition. In: Proceedings of the IEEE Conference on Computer Vision and Pattern Recognition, pp. 770–778 (2016)
7. Hu, J., Shen, L., Sun, G.: Squeeze-and-excitation networks. In: Proceedings of the IEEE Conference on Computer Vision and Pattern Recognition, pp. 7132–7141 (2018)

8. Huang, G., Liu, Z., Van Der Maaten, L., Weinberger, K.Q.: Densely connected convolutional networks. In: Proceedings of the IEEE Conference on Computer Vision and Pattern Recognition, pp. 4700–4708 (2017)

9. Huang, J., Ling, C.X.: Using AUC and accuracy in evaluating learning algorithms. IEEE Trans. Knowl. Data Eng. 17(3), 299–310 (2005)

10. Ioffe, S., Szegedy, C.: Batch normalization: accelerating deep network training by reducing internal covariate shift. arXiv preprint arXiv:1502.03167 (2015)

11. Kingma, D.P., Ba, J.: Adam: a method for stochastic optimization. arXiv preprint arXiv:1412.6980 (2014)

12. Krizhevsky, A., Sutskever, I., Hinton, G.E.: ImageNet classification with deep convolutional neural networks. Commun. ACM 60(6), 84–90 (2017)

13. Masood, A., Ali Al-Jumaily, A.: Computer aided diagnostic support system for skin cancer: a review of techniques and algorithms. Int. J. Biomed. Imaging 2013 (2013). https://www.hindawi.com/journals/ijbi/2013/323268/

14. Nadipineni, H.: Method to classify skin lesions using dermoscopic images. arXiv preprint arXiv:2008.09418 (2020)

15. Perez, F., Vasconcelos, C., Avila, S., Valle, E.: Data augmentation for skin lesion analysis. In: Stoyanov, D., et al. (eds.) CARE/CLIP/OR 2.0/ISIC -2018. LNCS, vol. 11041, pp. 303–311. Springer, Cham (2018). https://doi.org/10.1007/978-3-030-01201-4_33

16. Rezvantalab, A., Safigholi, H., Karimijeshni, S.: Dermatologist level dermoscopy skin cancer classification using different deep learning convolutional neural networks algorithms. arXiv preprint arXiv:1810.10348 (2018)

17. Selvaraju, R.R., Cogswell, M., Das, A., Vedantam, R., Parikh, D., Batra, D.: Grad-CAM: visual explanations from deep networks via gradient-based localization. In: Proceedings of the IEEE International Conference on Computer Vision, pp. 618–626 (2017)

18. Shaikh, M.A., Duan, T., Chauhan, M., Srihari, S.N.: Attention based writer independent verification. In: 2020 17th International Conference on Frontiers in Handwriting Recognition (ICFHR), September 2020. https://doi.org/10.1109/icfhr2020.2020.00074

19. Shen, S., et al.: Low-cost and high-performance data augmentation for deep-learning-based skin lesion classification. arXiv preprint arXiv:2101.02353 (2021)

20. Simonyan, K., Zisserman, A.: Very deep convolutional networks for large-scale image recognition. arXiv preprint arXiv:1409.1556 (2014)

21. Srivastava, N., Hinton, G., Krizhevsky, A., Sutskever, I., Salakhutdinov, R.: Dropout: a simple way to prevent neural networks from overfitting. J. Mach. Learn. Res. 15(1), 1929–1958 (2014)

22. Szegedy, C., Ioffe, S., Vanhoucke, V., Alemi, A.: Inception-v4, Inception-ResNet and the impact of residual connections on learning. arXiv preprint arXiv:1602.07261 (2016)

23. Tomita, N., Abdollahi, B., Wei, J., Ren, B., Suriawinata, A., Hassanpour, S.: Attention-based deep neural networks for detection of cancerous and precancerous esophagus tissue on histopathological slides. JAMA Netw. Open 2(11), e1914645 (2019)

24. Tran, D., Bourdev, L., Fergus, R., Torresani, L., Paluri, M.: Learning spatiotemporal features with 3D convolutional networks. In: Proceedings of the IEEE International Conference on Computer Vision, pp. 4489–4497 (2015)

25. Tschandl, P., Rosendahl, C., Kittler, H.: The HAM10000 dataset, a large collection of multi-source dermatoscopic images of common pigmented skin lesions. Sci. Data 5(1), 1–9 (2018)

26. Valle, E., et al.: Data, depth, and design: learning reliable models for skin lesion analysis. Neurocomputing **383**, 303–313 (2020)
27. Wang, F., et al.: Residual attention network for image classification (2017)
28. Xu, K., et al.: Show, attend and tell: neural image caption generation with visual attention. In: International Conference on Machine Learning, pp. 2048–2057 (2015)
29. Yao, P., et al.: Single model deep learning on imbalanced small datasets for skin lesion classification. arXiv preprint arXiv:2102.01284 (2021)
30. Yu, L., Chen, H., Dou, Q., Qin, J., Heng, P.A.: Automated melanoma recognition in dermoscopy images via very deep residual networks. IEEE Trans. Med. Imaging **36**(4), 994–1004 (2016)
31. Zhang, J., Xie, Y., Xia, Y., Shen, C.: Attention residual learning for skin lesion classification. IEEE Trans. Med. Imaging **38**(9), 2092–2103 (2019)
32. Zunair, H., Hamza, A.B.: Melanoma detection using adversarial training and deep transfer learning. Phys. Med. Biol. **65**, 135005 (2020)

Deep Grading Based on Collective Artificial Intelligence for AD Diagnosis and Prognosis

Huy-Dung Nguyen$^{(\boxtimes)}$, Michaël Clément, Boris Mansencal, and Pierrick Coupé

Univ. Bordeaux, CNRS, Bordeaux INP, LaBRI, UMR 5800, 33400 Talence, France
huy-dung.nguyen@u-bordeaux.fr

Abstract. Accurate diagnosis and prognosis of Alzheimer's disease are crucial to develop new therapies and reduce the associated costs. Recently, with the advances of convolutional neural networks, methods have been proposed to automate these two tasks using structural MRI. However, these methods often suffer from lack of interpretability, generalization, and can be limited in terms of performance. In this paper, we propose a novel deep framework designed to overcome these limitations. Our framework consists of two stages. In the first stage, we propose a deep grading model to extract meaningful features. To enhance the robustness of these features against domain shift, we introduce an innovative collective artificial intelligence strategy for training and evaluating steps. In the second stage, we use a graph convolutional neural network to better capture AD signatures. Our experiments based on 2074 subjects show the competitive performance of our deep framework compared to state-of-the-art methods on different datasets for both AD diagnosis and prognosis.

Keywords: Deep grading · Collective artificial intelligence · Generalization · Alzheimer's disease classification · Mild cognitive impairment

1 Introduction

The first cognitive symptoms of Alzheimer's disease (AD) appear right after the morphological changes caused by brain atrophy [10]. Those changes can be identified with the help of structural magnetic resonance imaging (sMRI) [2]. Recently, with the advances of convolutional neural networks (CNN), methods have been proposed for automatic AD diagnosis using sMRI. Despite encouraging results, current deep learning methods suffer from several limitations. First, deep models lack transparency in their decision-making process [31,38]. Therefore, this limits their use for computer-aided diagnosis tools in clinical practice. Second, for medical applications, the generalization capacity of classification models is essential. However, only a few works have proposed methods robust

© Springer Nature Switzerland AG 2021
M. Reyes et al. (Eds.): iMIMIC 2021/TDA4MedicalData 2021, LNCS 12929, pp. 24–33, 2021.
https://doi.org/10.1007/978-3-030-87444-5_3

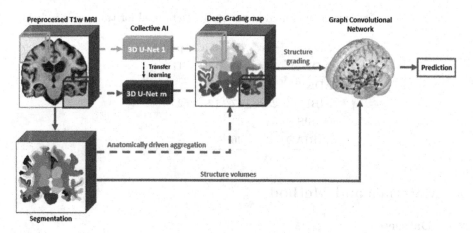

Fig. 1. Overview of our processing pipeline. The MRI image, its segmentation and the deep grading map illustrated are from an AD subject.

to domain shift [13,34]. Third, current CNN models proposed for AD diagnosis and prognosis still perform poorly [35]. Indeed, when properly validated on external datasets, current CNN-based methods perform worse than traditional approaches (*i.e.*, standard linear SVM).

In this paper, to address these three major limitations, we propose a novel interpretable, generalizable and accurate deep framework. An overview of our proposed pipeline is shown in Fig. 1. First, we propose a novel Deep Grading (DG) biomarker to improve the interpretability of deep model outputs. Inspired by the patch-based grading frameworks [4,12,32], this new biomarker provides a grading map with a score between −1 and 1 at each voxel related to the alteration severity. This interpretable biomarker may help clinicians in their decision and to improve our knowledge on AD progression over the brain. Second, we propose an innovative collective artificial intelligence strategy to improve the generalization across domains and to unseen tasks. As recently shown for segmentation [6,18], the use of a large number of networks capable of communicating offers a better capacity for generalization. Based on a large number of CNNs (*i.e.*, 125 U-Nets), we propose a framework using collective artificial intelligence efficient on different datasets and able to provide accurate prognosis while trained for diagnosis task. Finally, we propose to use a graph-based modeling to better capture AD signature using both inter-subject similarity and intrasubject variability. As shown in [12], such strategy improves performance in AD diagnosis and prognosis.

In this paper, our main contributions are threefold:

- A novel deep grading biomarker providing interpretable grading maps.
- An innovative collective artificial intelligence strategy robust to unseen datasets and unknown tasks.
- A new graph convolutional network (GCN) model for classification offering state-of-the-art performance for both AD diagnosis and prognosis.

Table 1. Number of participants used in our study. Data used for training is in bold.

Dataset	CN	AD	sMCI	pMCI
ADNI1	**170**	**170**	129	171
ADNI2	149	149	–	–
AIBL	233	47	12	20
OASIS3	658	97	–	–
MIRIAD	23	46	–	–

2   Materials and Method

2.1   Datasets

The data used in this study, consisting of 2074 subjects, were obtained from multiple cohorts: the Alzheimer's Disease Neuroimaging Initiative (ADNI) [16], the Open Access Series of Imaging Studies (OASIS) [21], the Australian Imaging, Biomarkers and Lifestyle (AIBL) [7], the Minimal Interval Resonance Imaging in Alzheimer's Disease (MIRIAD) [27]. We used the baseline T1-weighted MRI available in each of these studies. Each dataset contains AD patients and cognitively normal (CN) subjects. For ADNI1 and AIBL, it also includes mild cognitive impairment (MCI), the early stage of AD composed of abnormal memory dysfunctions. Two groups of MCI are considered: progressive MCI (pMCI) and stable MCI (sMCI). The definition of these two groups is the same as in [35]. Table 1 summarizes the number of participants for each dataset used in this study. During experiments, AD and CN from ADNI1 are used as training set and the other subjects as testing set.

2.2   Preprocessing

All the T1w MRI are preprocessed using the following steps: (1) denoising [29], (2) inhomogeneity correction [33], (3) affine registration into MNI space ($181 \times 217 \times 181$ voxels at $1\,mm \times 1\,mm \times 1\,mm$) [1], (4) intensity standardization [28] and (5) intracranial cavity (ICC) extraction [30]. After that preprocessing, we use AssemblyNet [6] to segment 133 brain structures (see Fig. 1). The list of structures is the same as in [14]. In this study, brain structure segmentation is used to determine the structure volume (*i.e.*, normalized volume in % of ICC) and aggregate information in the grading map (see Sect. 2.3 and Fig. 1).

2.3   Deep Grading for Disease Visualization

In AD classification, most of deep learning models only use CNN as binary classification tool. In this study, we propose to use CNN to produce 3D maps indicating where specific anatomical patterns are present and the importance of structural changes caused by AD.

To capture these anatomical alterations, we extend the idea of the patch-based grading (PBG) framework [4,12,32]. The PBG framework provides a 3D grading map with a score between −1 and 1 at each voxel related to the alteration severity. Contrary to previous PBG methods based on non-local mean strategy, here we propose a novel DG framework based on 3D U-Nets.

Concretely, each U-Net (similar to [6]) takes a 3D sMRI patch (e.g., 32 × 48 × 32) and outputs a grading map with values in range [−1, 1] for each voxel. Voxels with a higher value are considered closer to AD, while voxels with a lower value are considered closer to CN. For the ground-truth used during training, we assign the value 1 (resp. −1) to all voxels inside a patch extracted from an AD patient (resp. CN subject). All voxels outside of ICC are set to 0.

Once trained, the deep models are used to grade patches. These local outputs are gathered to reconstruct the final grading map (see Sect. 2.4). Using the structure segmentation, we represent each brain structure grading by its average grading score (see Fig. 1). This anatomically driven aggregation allows better and meaningful visualization of the disease progression. In this way, during the classification step (see Sect. 2.5), each subject is encoded by an n-dimensional vector where n is the number of brain structures.

2.4   Collective AI for Grading

As recently shown in [3,35], current AD classification techniques suffer from a lack of generalization. In this work, we propose an innovative collective artificial intelligence strategy to improve the generalization across domains and to unseen tasks. As recently shown for segmentation [6,18], the use of a large number of compact networks capable of communicating offers a better capacity for generalization. There are many advantages to using the collective AI strategy. First, it addresses the problem of GPU memory in 3D since each model processes only a sub-volume of the image. The use of a large number of compact networks is equivalent to a big neural network with more filters. Second, the voting system based on a large number of specialized and diversified models helps the final grading decision to be more robust against domain shift and different tasks.

Concretely, a preprocessed sMRI is decomposed into $k \times k \times k$ overlapping patches of the same size (e.g., 32 × 48 × 32). During training, for each patch localization in the MNI space, a specialized model is trained. Therefore, in our case ($k = 5$), we trained $m = k \times k \times k = 125$ U-Nets to cover the whole image (see Fig. 1). Moreover, each U-Net is initialized using transfer learning from its nearest neighbor U-Nets in the MNI space, except the first one trained from scratch as proposed in [6]. As adjacent patches have some common patterns, this communication allows grading models to share useful knowledge between them. For each patch, 80% of the training dataset (i.e., ADNI1) is used for training and the remaining 20% for validation. The accuracy obtained on validation set is used to reconstruct the final grading map using a weighted average as follows:

$$G_i = \frac{\sum_{x_i \in P_j} \alpha_j * g_{ij}}{\sum_{x_i \in P_j} \alpha_j} \tag{1}$$

where G_i is the grading score of the voxel x_i in the final grading map, g_{ij} is the grading score of the voxel x_i in the local grading patch P_j, and α_j is the validation accuracy of the patch j. This weighted vote enables to give more weight to the decision of accurate models during the reconstruction.

2.5   Graph Convolutional Neural Network for Classification

The DG feature provides an inter-subject similarity biomarker which is helpful to detect AD signature. However, the structural alterations leading to cognitive decline could be different between subjects. Indeed, following the idea of [12], we model the intra-subject variabilities by a graph representation to capture the relationships between several regions related to the disease. We define an undirected graph $G = (N, E)$, where $N = \{n_1, \dots, n_s\}$ is the set of nodes for the s brain structures and $E = s \times s$ is the matrix of edge connections. In our approach, all nodes are connected with each other in a complete graph, where nodes embed brain features (e.g., our proposed DG feature) and potentially other types of external features.

Indeed, besides the grading map, the volume of structures obtained from the segmentation could be helpful to distinguish AD patients from CN [12, 32]. It is due to the evidence that AD leads to structure atrophy. Age is also an important factor as, within sMRI, patterns in the brain of young AD patients could be similar to elder CN. Indeed, the combination of those features is expected to improve our classification performance. In our method, each node represents a brain structure and embeds a feature vector (DG, V, A) where V and A are respectively the volume of structures and subject's age. Finally, we use the graph convolutional neural network (GCN) [20] as the way to pass messages between nodes and to perform final classification.

2.6   Implementation Details

First, we downsample the sMRI from $181 \times 217 \times 181$ voxels (at $1\,\mathrm{mm}$) to $91 \times 109 \times 91$ voxels to reduce the computational cost, then decompose them into $5 \times 5 \times 5$ overlapping patches of size $32 \times 48 \times 32$ voxels equally spaced along the three axis. For each patch, an U-Net is trained using mean absolute error loss, Adam optimizer with a learning rate of 0.001. The training process is stopped after 20 epochs without improvement in validation loss. We employed several data augmentation and sampling strategies to alleviate the overfitting issue during training. A small perturbation is first created in training samples by randomly translating by $t \in \{-1, 0, 1\}$ voxel in 3 dimensions of the image. We then apply the mixup [37] data augmentation scheme that was shown to improve the generalization capacity of CNN in image classification.

Once the DG feature is obtained, we represent each subject by a graph of 133 nodes. Each node represents a brain structure and embeds DG, volume and age features. Our classifier is composed of 3 layers of GCN with 32 channels, followed by a global mean average pooling layer and a fully connected layer with an output size of 1. The model is trained using the binary cross-entropy loss, Adam optimizer with a learning rate of 0.0003. The training process is stopped after 20 epochs without improvement in validation loss. At inference time, we randomly add noise $X \sim \mathcal{N}(0, 0.01)$ to the node features and compute the average of 3 predictions to get the global decision. Experiments have shown that it helps our GCN to be more stable.

For training and evaluating steps, we use a standard GPU (*i.e.*, NVIDIA TITAN X) with 12 Gb of memory.

3   Experimental Results

In this study, the grading models and classifiers are trained using ADNI1 dataset within AD and CN subjects. Then, we assess their generalization capacity in domain shift using AD, CN subjects from ADNI2, AIBL, OASIS, MIRIAD. The generalization capacity in derived tasks is performed using pMCI, sMCI subjects from ADNI1 (same domain) and AIBL (out of domain).

Influence of Collective AI Strategy. In this part, the DG feature is denoted as DG_C(resp. DG_I) when obtained with the collective (resp. individual) AI strategy. The individual AI strategy refers to the use of a single U-Net to learn patterns from all patches of sMRI. We compare the efficiency of DG_C and DG_I feature when using the same classifier (*i.e.*, SVM or GCN) (see Table 2). These experiments show that using DG_C achieves better results in most configurations. When using SVM classifier, we observe a gain of 3.6% (resp. 0.8%) on average in AD/CN (resp. pMCI/sMCI) classification. The efficiency of G_C feature is even better with GCN classifier, where a gain of 4.0% (resp. 3.5%) is observed.

Influence of GCN Classifier. Besides the DG feature, the intra-subject variabilities are also integrated into our graph representation. Hence, it should be beneficial to use GCN to exploit all this information. In our experiments, GCN outperforms SVM in all the tests using either DG_I or DG_C feature (see Table 2). Concretely, using DG_I feature, we observe a gain of 5.0% (resp. 7.6%) on average for AD/CN (resp. pMCI/sMCI) classification. These improvements are 5.4% and 10.6% when using DG_C feature.

Influence of Using Additional Non-image Features. Moreover, we analyze the model performance using DG_C with the structural volume V and age A as additional node features in our graph representation. By using the combined features, the performance on average in AD/CN and pMCI/sMCI is both improved by 0.3% and 1.4% compared to DG_C feature (see Table 2). In the rest of this paper, these results are used to compare with current methods.

Comparison with State-of-the-Art Methods. Table 3 summarizes the current performance of state-of-the-art methods proposed for AD diagnosis and prognosis classification that have been validated on external datasets. In this comparison we considered five categories of deep methods: patch-based strategy based on a single model (Patch-based CNN [35]), patch-based strategy based on multiple models (Landmark-based CNN [24], Hierarchical FCN [23]), ROI-based

Table 2. Validation of the collective AI strategy, GCN classifier, the combination of DG feature with other image and non-image features using GCN classifier. Red: best result, Blue: second best result (see online version for colors). The balanced accuracy (BACC) is used to assess the model performance. The results are the average accuracy of 10 repetitions and presented in percentage. All the methods are trained on the AD/CN subjects of the ADNI1 dataset.

Classifier	Features	AD/CN				pMCI/sMCI		Average	
		ADNI2	OASIS	MIRIAD	AIBL	ADNI1	AIBL	AD/CN	p/sMCI
SVM	DG_I	83	83	88	79	65	66	83.3	65.5
SVM	DG_C	83	84	91	87	68	64	86.3	66.0
GCN	DG_I	84	88	96	82	68	73	87.5	70.5
GCN	DG_C	87	89	100	88	70	76	91.0	73.0
GCN	DG_C, V, A	87	88	98	92	74	74	91.3	74.0

Table 3. Comparison of our method with current methods in AD diagnosis and prognosis. Red: best result, Blue: second best result (see online version for colors). The balanced accuracy (BACC) is used to assess the model performance. All the methods are trained on the AD/CN subject of the ADNI1 dataset (except [23] that is fined-tuned on MCI subjects for sMCI/pMCI task).

Methods	AD/CN				pMCI/sMCI	
	ADNI2	OASIS	MIRIAD	AIBL	ADNI1	AIBL
Landmark-based CNN [24]	91	–	92	–	–	–
Hierarchical FCN [23]	89	–	–	–	69	–
Patch-based CNN [35]	–	64	–	81	70	64
ROI-based CNN [35]	–	69	–	84	70	60
Subject-based CNN [35]	–	67	–	83	69	52
Voxel-based SVM [35]	–	70	–	88	75	62
AD^2A [11]	88	–	–	88	–	–
Efficient 3D [36]	–	92	96	91	70	65
3D Inception-ResNet-v2 [26]	–	85	–	91	42	–
Our method	87	88	98	92	74	74

strategy based on a single model focused on hippocampus (ROI-based CNN [35]), subject-based considering the whole image based on a single model (subject-based CNN [35], 3D Inception-ResNet-v2 [26], Efficient 3D [36] and AD^2A [11]) and a classical voxel-based model using a SVM (Voxel-based SVM [35]).

For AD diagnosis (*i.e.*, AD/CN), all the methods show good balanced accuracy, although some of them failed to generalize on OASIS. In this scenario (unseen datasets), our method obtained high accuracy for all the datasets. This confirms the generalization capacity of our approach against domain shift.

For AD prognosis (*i.e.*, pMCI/sMCI), we observe a significant drop for all the methods. This drop is expected since pMCI/sMCI classification is more challenging and since models are trained on a different task (*i.e.*, AD/CN). For this task, our method is generally robust, especially on AIBL. Moreover, our approach is the only deep learning method that performs competitively with the SVM model [35] on ADNI1, while significantly better on AIBL. In this scenario (unknown task), our method obtains the highest accuracy on average. These results highlight the potential performance of our method on unseen tasks.

Interpretation of Collective Deep Grading. To highlight the interpretability capabilities offered by our DG feature, we first compute the average DG map for each group: AD, pMCI, sMCI and CN (see Fig. 2). First, we can note that the average grading maps increase between each stage of the disease. Second, we estimated the top 10 structures with highest absolute value of grading score over all the testing subjects. The found structures are known to be specifically and early impacted by AD. These structures are: *bilateral hippocampus* [9], *left amygdala* and *left inferior lateral ventricle* [5], *left parahippocampal gyrus* [19], *left posterior insula* [8], *left thalamus proper* [17], *left transverse temporal gyrus* [25], *left ventral diencephalon* [22]. While other attention-based deep methods failed to find structures related to AD [3], our DG framework shows high correlation with current physiopathological knowledge on AD [15].

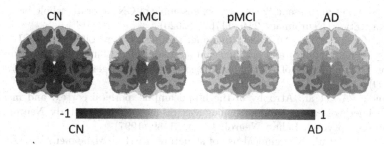

Fig. 2. Average grading map per group of subjects.

4 Conclusion

In this paper, we addressed three major limitations of CNN-based methods by introducing a novel interpretable, generalizable and accurate deep grading framework. First, deep grading offers a meaningful visualization of the disease progression. Second, we proposed a collective artificial intelligence strategy to improve the generalization of our DG strategy. Experimental results showed a gain for both SVM and GCN in all tasks using this strategy. Finally, we proposed to use a graph-based modeling to better capture AD signature using both inter-subject similarity and intra-subject variability. Based on that, our DG method showed state-of-the-art performance in both AD diagnosis and prognosis.

Acknowledgments. This work benefited from the support of the project Deepvol-Brain of the French National Research Agency (ANR-18-CE45-0013). This study was achieved within the context of the Laboratory of Excellence TRAIL ANR-10-LABX-57 for the BigDataBrain project. Moreover, we thank the Investments for the future Program IdEx Bordeaux (ANR-10-IDEX-03-02), the French Ministry of Education and Research, and the CNRS for DeepMultiBrain project.

References

1. Avants, B.B., et al.: A reproducible evaluation of ANTs similarity metric performance in brain image registration. Neuroimage **54**(3), 2033–2044 (2011)
2. Bron, E., et al.: Standardized evaluation of algorithms for computer-aided diagnosis of dementia based on structural MRI: the CADDementia challenge. Neuroimage **111**, 562–579 (2015)
3. Bron, E., et al.: Cross-cohort generalizability of deep and conventional machine learning for MRI-based diagnosis and prediction of Alzheimer's disease. NeuroImage Clin. **31**, 102712 (2021)
4. Coupé, P., et al.: Scoring by nonlocal image patch estimator for early detection of Alzheimer's disease. NeuroImage Clin. **1**(1), 141–152 (2012)
5. Coupé, P., et al.: Lifespan changes of the human brain in Alzheimer's disease. Nat. Sci. Rep. **9** (2019). Article number: 3998. https://www.nature.com/articles/s41598-019-39809-8
6. Coupé, P., et al.: AssemblyNet: a large ensemble of CNNs for 3D whole brain MRI segmentation. Neuroimage **219**, 117026 (2020)
7. Ellis, K., et al.: The Australian Imaging, Biomarkers and Lifestyle (AIBL) study of aging: methodology and baseline characteristics of 1112 individuals recruited for a longitudinal study of Alzheimer's disease. Int. Psychogeriatr. IPA **21**, 672–687 (2009)
8. Foundas, A., et al.: Atrophy of the hippocampus, parietal cortex, and insula in Alzheimer's disease: a volumetric magnetic resonance imaging study. Neuropsychiatry Neuropsychol. Behav. Neurol. **10**(2), 81–89 (1997)
9. Frisoni, G., et al.: The clinical use of structural MRI in Alzheimer's disease. Nat. Rev. Neurol. **6**, 67–77 (2010)
10. Gordon, B., et al.: Spatial patterns of neuroimaging biomarker change in individuals from families with autosomal dominant Alzheimer's disease: a longitudinal study. Lancet Neurol. **17**, 241–250 (2018)
11. Guan, H., Yang, E., Yap, P.-T., Shen, D., Liu, M.: Attention-guided deep domain adaptation for brain dementia identification with multi-site neuroimaging data. In: Albarqouni, S., et al. (eds.) DART/DCL-2020. LNCS, vol. 12444, pp. 31–40. Springer, Cham (2020). https://doi.org/10.1007/978-3-030-60548-3_4
12. Hett, K., Ta, V.-T., Manjón, J.V., Coupé, P.: Graph of brain structures grading for early detection of Alzheimer's disease. In: Frangi, A.F., Schnabel, J.A., Davatzikos, C., Alberola-López, C., Fichtinger, G. (eds.) MICCAI 2018. LNCS, vol. 11072, pp. 429–436. Springer, Cham (2018). https://doi.org/10.1007/978-3-030-00931-1_49
13. Hosseini-Asl, E., et al.: Alzheimer's disease diagnostics by adaptation of 3D convolutional network. In: IEEE International Conference on Image Processing (ICIP) (2016)
14. Huo, Y., et al.: 3D whole brain segmentation using spatially localized atlas network tiles. Neuroimage **194**, 105–119 (2019)
15. Jack, C., et al.: Hypothetical model of dynamic biomarkers of the Alzheimer's pathological cascade. Lancet Neurol. **27**, 685–691 (2010)
16. Jack, C.R., Jr., et al.: The Alzheimer's disease neuroimaging initiative (ADNI): MRI methods. J. Magn. Reson. Imaging **27**(4), 685–691 (2008)

17. de Jong, L.W., et al.: Strongly reduced volumes of putamen and thalamus in Alzheimer's disease: an MRI study. Brain **131**(12), 3277–3285 (2008)
18. Kamraoui, R.A., et al.: Towards broader generalization of deep learning methods for multiple sclerosis lesion segmentation. arXiv arXiv:2012.07950 (2020)
19. Kesslak, J.P., et al.: Quantification of magnetic resonance scans for hippocampal and parahippocampal atrophy in Alzheimer's disease. Neurology **41**(1), 51 (1991)
20. Kipf, T., et al.: Semi-supervised classification with graph convolutional networks. In: International Conference on Learning Representations (ICLR) (2017)
21. LaMontagne, P.J., et al.: OASIS-3: longitudinal neuroimaging, clinical, and cognitive dataset for normal aging and Alzheimer disease. MedRxiv (2019)
22. Lebedeva, A.K., et al.: MRI-based classification models in prediction of mild cognitive impairment and dementia in late-life depression. Front. Aging Neurosci. **9**, 13 (2017)
23. Lian, C., et al.: Hierarchical fully convolutional network for joint atrophy localization and Alzheimer's disease diagnosis using structural MRI. IEEE Trans. Pattern Anal. Mach. Intell. **42**(4), 880–893 (2020)
24. Liu, M., et al.: Landmark-based deep multi-instance learning for brain disease diagnosis. Med. Image Anal. **43**, 157–168 (2017)
25. Liu, Y., et al.: Education increases reserve against Alzheimer's disease-evidence from structural MRI analysis. Neuroradiology **54**, 929–938 (2012). https://doi.org/10.1007/s00234-012-1005-0
26. Lu, B., et al.: A practical Alzheimer disease classifier via brain imaging-based deep learning on 85,721 samples. BioRxiv (2021)
27. Malone, I.B., et al.: MIRIAD–public release of a multiple time point Alzheimer's MR imaging dataset. Neuroimage **70**, 33–36 (2013)
28. Manjón, J.V., et al.: Robust MRI brain tissue parameter estimation by multistage outlier rejection. Magn. Reson. Med. **59**(4), 866–873 (2008)
29. Manjón, J.V., et al.: Adaptive non-local means denoising of MR images with spatially varying noise levels. J. Magn. Reson. Imaging **31**(1), 192–203 (2010)
30. Manjón, J.V., et al.: NICE: non-local intracranial cavity extraction. Int. J. Biomed. Imaging (2014)
31. Nigri, E., et al.: Explainable deep CNNs for MRI-based diagnosis of Alzheimer's disease. In: International Joint Conference on Neural Networks (IJCNN) (2020)
32. Tong, T., et al.: A novel grading biomarker for the prediction of conversion from mild cognitive impairment to Alzheimer's disease. IEEE Trans. Biomed. Eng. **64**(1), 155–165 (2016)
33. Tustison, N.J., et al.: N4ITK: improved N3 bias correction. IEEE Trans. Med. Imaging **29**(6), 1310–1320 (2010)
34. Wachinger, C., et al.: Domain adaptation for Alzheimer's disease diagnostics. Neuroimage **139**, 470–479 (2016)
35. Wen, J., et al.: Convolutional neural networks for classification of Alzheimer's disease: overview and reproducible evaluation. Med. Image Anal. **63**, 101694 (2020)
36. Yee, E., et al.: Construction of MRI-based Alzheimer's disease score based on efficient 3D convolutional neural network: comprehensive validation on 7,902 images from a multi-center dataset. J. Alzheimers Dis. **79**, 1–12 (2020)
37. Zhang, H., et al.: mixup: beyond empirical risk minimization. In: International Conference on Learning Representations (ICLR) (2018)
38. Zhang, X., et al.: An explainable 3D residual self-attention deep neural network for joint atrophy localization and Alzheimer's disease diagnosis using structural MRI. IEEE J. Biomed. Health Inform. (2021)

This Explains *That*: Congruent Image–Report Generation for Explainable Medical Image Analysis with Cyclic Generative Adversarial Networks

Abhineet Pandey[1], Bhawna Paliwal[1], Abhinav Dhall[1,2],
Ramanathan Subramanian[1,3], and Dwarikanath Mahapatra[4(✉)]

[1] Indian Institute of Technology (IIT) Ropar, Ropar, India
[2] Monash University, Melbourne, Australia
[3] University of Canberra, Canberra, Australia
[4] Inception Institute of Artificial Intelligence, Abu Dhabi, United Arab Emirates
dwarikanath.mahapatra@inceptioniai.org

Abstract. We present a novel framework for *explainable labeling and interpretation* of medical images. Medical images require specialized professionals for interpretation, and are explained (typically) via elaborate textual reports. Different from prior methods that focus on medical report generation from images or vice-versa, we novelly generate congruent image–report pairs employing a cyclic-Generative Adversarial Network (cycleGAN); thereby, the generated report will adequately explain a medical image, while a report-generated image that effectively characterizes the text visually should (sufficiently) resemble the original. The aim of the work is to generate *trustworthy and faithful explanations* for the outputs of a model diagnosing chest X-ray images by pointing a human user to similar cases in support of a diagnostic decision. Apart from enabling transparent medical image labeling and interpretation, we achieve report and image-based labeling comparable to prior methods, including state-of-the-art performance in some cases as evidenced by experiments on the *Indiana Chest X-ray* dataset.

Keywords: Explainability · Medical image analysis · Multimodal

1 Introduction

Medical images present critical information for clinicians and epidemiologists to diagnose and treat a variety of diseases. However, unlike natural images (scenes) which can be easily analyzed and explained by laypersons, medical images are hard to understand and interpret without specialized expertise [1].

Artificial intelligence (AI) has made rapid advances in the last decade thanks to deep learning. However, the need for accountability and transparency to

A. Pandey and B. Paliwal—contributed equally.

© Springer Nature Switzerland AG 2021
M. Reyes et al. (Eds.): iMIMIC 2021/TDA4MedicalData 2021, LNCS 12929, pp. 34–43, 2021.
https://doi.org/10.1007/978-3-030-87444-5_4

explain decisions along with high performance, especially in healthcare, has spurred the need for *explainable machine learning*. While natural images can be analyzed and explained by *decomposing* them into semantically-consistent and prototypical visual segments citech4thislooks, multimodal approaches for prototypical explanations are essential for interpreting and explaining medical imagery given the tight connection between image and text in this domain.

1.1 Prior Work

Prior works on medical image interpretation and explainability have either attempted to characterize (chest) x-rays in terms of multiple pathological labels [19] or via automated generation of imaging reports [1,11,15]. The Chexnet framework [19] employs a 121-layer convolution network to label chest x-rays. A multi-task learning framework is employed to generate both tags and elaborate reports via a hierarchical long short-term memory (LSTM) model in [1]. Improvements over [1] are achieved by [11] and [15] by employing a topic model and a memory driven transformer respectively. While the above report-generation works achieve excellent performance, and effectively learn mappings between the image and textual features, they nevertheless do not *verify* if the generated report characterizes the input x-ray. It is this constrained characterization in our suggested work that helps us generate prototypical chest x-ray images serving as explanations. In more recent work saliency maps have been used to select informative xray images [16].

1.2 Our Approach

This work differently focuses on the generation of *coherent* image–report pairs, and posits that if the image and report are conjoined counterparts, one should inherently describe the characteristics of the other. It is the second part of the radiology report generation model i.e. generation of prototypical images from generated reports that serve as explanations for the generated reports. The explainable model proposed can be characterised as a model having post hoc explanations where an explainer outputs the explanations corresponding to the output of the model being explained. The approach to explanations in such an explanation technique as ours is different from methods which propose simpler models such as decision trees that are inherently explainable. Having prototypical images as explanations has been used in case of natural images in [18] (discussed earlier) and [26]. None of the approaches explores the paradigm of prototypical image generation as explanations in case of medical images which has been proposed in this work novelly with a multimodal approach.

1.3   Contributions

Overall, we make the following research contributions:

1. We present the first multimodal formulation that enforces the generation of *coherent* and *explanatory* image–report pairs via the cycle-consistency loss employed in cycleGANs [14].
2. Different from prior works, we regenerate an x-ray image from the report, and use this image to quantitatively and qualitatively evaluate the report quality. Extensive labeling experiments on textual reports and images generated via the Indiana Chest X-ray dataset [20] reveal the effectiveness of our multi modal explanations approach.
3. We evaluate the proposed model on two grounds namely: the quality of generated reports and the quality of generated explanations. Our method achieves results comparable to prior methods in report generation task, while achieving state-of-the-art performance in certain conditions. The evaluations done for post-hoc explanations show the employability of cycle consistency constraints and multimodal analysis as an explanation technique.
4. As qualitative evaluation, we present Grad CAM [4]-based *attention maps* conveying where a classification model *focuses* to make a prediction.

2   Method

2.1   Coherent Image-Report Pairs with CycleGANs

We aim to model the tight coherence between image and textual features in the chest x-ray images and reports through our multi-modal report generation model. Reports generated should be such that an x-ray image generated with just these generated reports as the input should be similar to ground truth x-ray images; and prototypical x-ray images generated as explanations should be such that a report generated from these images as inputs resembles original report. We hence devise a multimodal, paired GAN architecture explicitly modeling the cycle consistent constraints based on CycleGAN [14] with data of type {*image, text/labels*}.

2.2   CycleGAN

Given two sets of images corresponding to domains X and Y (for example, two sets of paintings corresponding to different styles), cycleGAN enables learning a mapping $G: X \rightarrow Y$ such that the generated image $G(x) = y'$, where $x \in X$ and $y \in Y$, looks similar to y.

The generated images y' are also mapped back to images x' in domain X. Hence, cycleGAN also learns another mapping $F: Y \rightarrow X$ where $F(y') = x'$ such that x' is similar to x. The structural cycle-consistency assumption is modeled via the cycle consistency loss, which enforces $F(G(x))$ to be similar to x, and

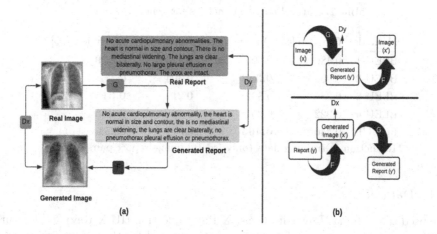

Fig. 1. a) Representation of our multimodal cycleGAN framework with exemplar inputs and generated outputs for the image and text modalities. b) Application of the cyclic GAN [14] framework to generate coherent image–report pairs.

conversely, $G(F(y))$ to be similar to y. Hence the objective loss to be minimized enforces the following four constraints:

$$G(x) \approx y, F(y) \approx x \quad and \quad F(G(x)) \approx x, G(F(x)) \approx y \qquad (1)$$

We exploit the setting of Cycle-GAN in a multimodal paradigm i.e. the domains in which we work are text (reports) and image (chest x-ray). As shown in Fig. 1, our multimodal cyclic GAN architecture comprises (i) two GANs F and G to respectively generate images from textual descriptions and vice-versa, and (ii) two deep neural networks, termed *Discriminators* D_X and D_Y, to respectively compare the generated images and reports against the originals. Figure 1(a) depicts the mappings G and F, while Fig. 1(b) depicts how cycle-consistency is enforced to generate coherent image-report pairs.

2.3   Explanatory Image–Report Pairs

Our model learns mappings between prototypical image–text decompositions (termed visual or textual words in information retrieval) akin to the *this looks like that* formulation [18] and synthetic image based explanations in [26]. Since our setting is multimodal instead of image to image setting in cycle-gans, GAN G (report-to-image generator) in our setting is based on a CNN-plus-LSTM based generative model similar to the architecture proposed in [1]. GAN F (image-to-report generation) uses a hierarchical structure composed of two GANs similar to [10]. First, GAN F_1 takes the text embedding as input and generates a low-resolution (64×64) image. The second GAN F_2 utilizes this image and the text embedding to generate a high-resolution (256×256) image.

Table 1. Natural language metrics for generated reports

Methods:	Ours-cycle*	Ours-no-cycle	R2Gen [15]	Multiview [25]
BLEU-1	0.486	0.520	0.470	**0.529**
BLEU-2	0.360	**0.388**	0.304	0.372
BLEU-3	0.285	0.302	0.219	**0.315**
BLEU-4	0.222	0.251	0.165	**0.255**
ROUGE	0.440	**0.463**	0.371	0.453

*Reduction in training data (only frontal image-report pairs used)

2.4 Dataset

We used the Indiana University Chest X-Ray Collection (IU X-Ray) [20] for our experiments, as it contains free text reports essential for the report generation task. IU X-Ray is a set of chest x-ray images paired with their corresponding diagnostic reports. The dataset contains 7,470 images, some of which map to the same free text report. 51% of the images are frontal, while the other 49% are lateral.

The frontal and lateral images map to individual text reports, at times corresponding to the same report. Consequently, mapping reports to images may confound the generator F regarding which type of image to generate. To avoid this confusion, we work only with frontal images, thus reducing the dataset to 3793 image-text pairs. Each report consists of the following sections: impression, findings, tags, comparison, and indication. In this work, we treat the contents of impression and findings as the target captions to be generated. We adopt a 80:20 train-test split for all experiments.

2.5 Implementation

All images were resized to 244 × 224 size. We used 512 × 512 images for initial experiments involving the 'Ours-no-cycle' method (see Table 1) and observed a better performance with respect to natural language metrics. However, low-resolution x-rays were used for subsequent experiments due to computational constraints. The input and hidden state dimensions for Sentence-LSTM are 1024 and 512 respectively, while both are of 512 length in the case of Word-LSTM. Learning rate used for the visual encoder is 1e−5, while 5e−4 is used for LSTM parts. Embedding dimension used for input to the text-to-image framework is 256, with learning rate set to 2e−4 for both the discriminator and the generator. We used PyTorch [5]-based implementations for all experiments.

Firstly, we individually trained the image-to-text and text-to-image generator modules. In the text-to-image part, we first trained the Stage 1 generator, followed by Stage 2 training on freezing the Stage 1 generator. Note that this individual training of the text-to-image module was done on original reports from the training set. However, when we trained the cycleGAN architecture, the text-to-image part took in the generated text as input. While directly training both the modules together, oscillations in loss values were observed.

3 Evaluation

3.1 Evaluation of Generated Reports

We first evaluate the quality of the generated reports via the BLEU and ROUGE metrics [23,24]; we compare our performance against other methods [15,25] in Table 1. Our methods with and without cycle-consistency loss are referred to as *Ours-cycle* and *Ours-no-cycle*. Since only frontal images were used for training *Ours-cycle* (see Sect. 2.5), the training set is reduced to 3793 image–report pairs. We get comparable performance with the multi-view network [25] based on NLG metrics. There is a small drop in these metrics with the addition of the cycle component, mainly due to the reduction in training data (as the number of image-report repairs is approximately halved).

3.2 Evaluation of Explanations

To evaluate the explanations, we first assess if the generated images truly resemble real input images because the quality of the generated images is also a representative of the quality of the model generated reports as discussed in earlier sections. Secondly, we consider the aspects of trust and faithfulness of our explanation technique based on ideas in [27] for post-hoc explanations.

3.2.1 Evaluating Similarity of Generated Images and Real X-Ray Images

We quantitatively assess the images using CheXNet [19] (state-of-the-art performance on multi-label classification for chest x-ray images). We use CheXNet on ⟨input image–generated image⟩ pairs for checking the amount of disparity present between the *true* and *generated* images. We achieve a KL-Divergence of 0.101. We also introduce a 'top-k' metric to identify if the same set of diseases are identified from the *input* and *generated* images. The metric averages the number of top predicted diseases which are *common* to both input and the generated images.

$$top - k = \frac{\sum_{All\,pairs} |\,(top - k\,labels(input\,image)) \cap (top - k\,labels(generated\,image))\,|}{Number\,of\,pairs}$$

We compare the output labels of CheXNet on both real and generated image using the top-k, Precision@k and Recall@k metrics. From Table 2, on average 1.84 predicted disease labels are common between the input and generated images, considering only the top-two ranked disease labels. In Table 2, we have also shown a comparison against images generated from our text-to-image (report-to-x-ray-image) model on the reports generated by the recently proposed

Table 2. Metrics for generated images by using CheXNet for multi-label classification.

k	Top-k (Ours)	Top-k (R2Gen) [15]	Precision@k (Ours)	Precision@k (R2Gen) [15]	Recall@k (Ours)	Recall@k (R2Gen) [15]
2	**1.84**	0.64	**0.92**	0.32	**0.13**	0.05
5	**3.01**	2.55	**0.60**	0.51	**0.21**	0.18
8	**6.45**	5.82	**0.81**	0.73	**0.46**	0.42

Table 3. Accuracy metric for the reports generated from prototypical (generated) images

Label	No Finding	Cardiomediastinum	Cardiomegaly	Lung Lesion	Lung Opacity	Edema	Consolidation	Pneumonia	Atelectasis	Pneumothorax	Pleural(E)	Pleural(O)	Fracture	Support Devices
Accuracy	0.78	0.92	0.84	0.96	0.82	0.97	0.96	0.97	0.94	0.98	0.95	0.99	0.96	0.94

transformer-based R2gen algorithm [15]. Our representative generated images perform better on the top-x, precision and recall metrics, quantitatively showing that the reports generated by our cycleGAN model better describe the input chest x-ray image.

3.2.2 Evaluating Trustability of the Explanations

We build upon the idea of trust in an explanation technique suggested in [27] for post-hoc explanations. An explanation method can be considered trustworthy if the generated explanations are able to characterize the kind of inputs on which the model performs well or closer to the ground truth. We evaluate our explanations on this aspect of trustability by testing if the explanations or prototypical x-ray images generated are the images on which reports generated are very close to ground truth reports. We evaluate the similarity of the two reports (ground truth reports and reports generated from prototypical images) by comparing the labels output by a naive Bayes classifier on the input reports. The results for accuracy metric for each of the 14 labels is summarised in the Table 3. We can clearly infer that the x-ray images generated as explanations have been able to understand the model's behaviour and hence the good accuracy (around 0.9 for most of the labels).

3.2.3 Evaluating Faithfulness of Explanations

Another aspect which has been explored in some of the explanation works is faithfulness of the technique i.e. whether the explanation technique is reliable. Reliability is understood in the sense that it is reflecting the underlying associations of the model rather than any other correlation such as just testing the presence of edges in object detection tasks [12]. We test the faithfulness of the explanations generated by randomising the weights of the report generation model and then evaluating the quality of prototypical images to check if the explanation technique can be called faithful to the model parameters. The metric values for Top-2, Precision@2 and Recall@2 for generated images in this case are 0.90, 0.45 and 0.06 respectively significantly less than corresponding metrics in Table 2. As evident, the prototypical images generated as explanations from randomised weights model are unable to characterize the original input images because the model they are explaining doesn't contain the underlying information it had previously learnt for characterizing given chest x-ray images.

3.2.4 Qualitative Assessment of Generated Images Using Grad-CAM

We used GradCAM [4] for highlighting the salient image regions focused upon by the CheXNet [19] model for label prediction from the real and generated image pairs. Two examples are shown in Fig. 2. In the left sample pair, real image shows fibrosis as the most probable disease label, as does the generated image. As observable, the highlighted region showing the presence of a nodule is the same in both x-ray images except for the flip from the left and right lung. This shows that the report generation model was able to capture these abnormalities with great detail, as the report-generated image also captures these details visually. Similarly, two of the top-three labels are the same in both real and generated images as predicted by CheXNet in sample pair 2.

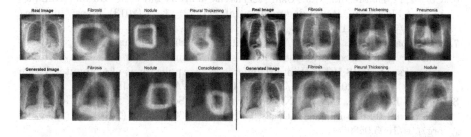

Fig. 2. Grad CAM saliency maps for top 3 predicted labels by CheXNet for real (top row) and generated (bottom row) image pairs; Sample pair 1 (left) and Sample pair 2 (right)

4   Conclusion

A cycleGAN-based framework for explainable medical report generation and synthesis of coherent image-report pairs is proposed in this work. Our generated images visually characterize the text reports, and resemble the input image with respect to pathological characteristics. We have performed extensive experiments and evaluation on the generated images and reports, which show that our report-generation quality is comparable to the state-of-the-art in terms of natural language generation metrics; also the generated images depict the disease attributes both via attention maps and other quantitative measures (precision analysis, trust, and faithfulness) showing the usefulness of a cycle-constrained characterization of chest x-ray images in an explainable medical image analysis task.

References

1. Jing, B., Xie, P., Xing, E.: On the automatic generation of medical imaging reports. In: Proceedings of the 56th Annual Meeting of the Association for Computational Linguistics (2018)
2. Girshicka, R., Donahuea, J., Darrell, T., Malik, J.: Rich feature hierarchies for accurate object detection and semantic segmentation. In: CVPR (2014)
3. Krizhevsky, A., Sutskever, I., Hinton, G.E.: ImageNet classification with deep convolutional neural networks. In: NIPS (2012)
4. Selvaraju, R.R., Das, A., Vedantam, R., Cogswell, M., Parikh, D., Batra, D.: Grad-CAM: why did you say that? visual explanations from deep networks via gradient-based localization. CoRR, abs/1610.02391 (2016). http://arxiv.org/abs/1610.02391
5. Paszke, A., et al.: Pytorch: an imperative style, high-performance deep learning library. In: Wallach, H., Larochelle, H., Beygelzimer, A., dAlche-Buc, F., Fox, E., Garnett, R. (eds.) Advances in Neural Information Processing Systems, vol. 32, pp. 8024–8035. Curran Associates Inc, (2019). http://papers.neurips.cc/paper/9015-pytorch:an-imperative-style-high-performance-deep-learning-library.pdf
6. Cortez, P., Embrechts, M.J.: Using sensitivity analysis and visualization techniques to open black box data mining models. Inf. Sci. **225**, 1–17 (2013). https://doi.org/10.1016/j.ins.2012.10.039
7. Chattopadhay, A., Sarkar, A., Howlader, P., Balasubramanian, V.N.: Grad-cam++: generalized gradient based visual explanations for deep convolutional networks. IN: 2018 IEEE Winter Conference on Applications of Computer Vision (WACV), March 2018. https://doi.org/10.1109/WACV.2018.00097
8. Reed, S.E., Akata, Z., Yan, X., Logeswaran, L., Schiele, B., Lee, H.: Generative adversarial text to image synthesis. CoRR, abs/1605.05396, (2016). http://arxiv.org/abs/1605.05396
9. Xu, T., et al.: Attngan: fine-grained text to image generation with attentional generative adversarial networks. CoRR, abs/1711.10485 (2017). http://arxiv.org/abs/1711.10485
10. Zhang, H., et al.: Stackgan: text to photo-realistic image synthesis with stacked generative adversarial networks. CoRR, vol. abs/1612.03242 (2016). http://arxiv.org/abs/1612.03242
11. Liu, G., et al.: Clinically accurate chest x-ray report generation. CoRR, abs/1904.02633 (2019). http://arxiv.org/abs/1904.02633
12. Adebayo, J., Gilmer, J., Muelly, M., Goodfellow, I., Hardt, M., Kim, B.: Sanity checks for saliency maps (2020)
13. Ribeiro, M.T., Singh, S., Guestrin, C.: Why should i trust you?: Explaining the predictions of any classifier (2016)
14. Zhu, J.Y., Park, T., Isola, P., Efros, A.A.: Unpaired image-to-image translation using cycle-consistent adversarial networks (2020)

15. Chen, Z., Song, Y., Chang, T.H., Wan, X.: Generating radiology reports via memory-driven transformer (2020). In: Mahapatra, D., Poellinger, A., Shao, L., Reyes, M. (eds.) @articleMahapatraTMI2021, Interpretability-Driven Sample Selection Using Self Supervised Learning For Disease Classification And Segmentation, pp. 1–15. IEEE (2021)

16. Mahapatra, D., Poellinger, A., Shao, L., Reyes, M.: A interpretability-driven sample selection using self supervised learning for disease classification and segmentation. IEEE Trans. Med. Imag. (2021)

17. Krause, J., Johnson, J., Krishna, R., Fei-Fei, L.: A hierarchical approach for generating descriptive image paragraphs. In: Computer Vision and Pattern Recognition (CVPR) (2017)

18. Chen, C., Li, D., Barnett, A., Su, J., Rudin, C.: This looks like that: deep learning for interpretable image recognition. CoRR, abs/1806.10574 (2018). http://arxiv.org/abs/1806.10574

19. Rajpurkar, P., et al.: Chexnet: radiologist-level pneumonia detection on chest x-rays with deep learning. CoRR, abs/1711.05225 (2017). http://arxiv.org/abs/1711.05225

20. Demner-Fushman, D., et al.: Preparing a collection of radiology examinations for distribution and retrieval. J. Am. Med. Inform. Assoc. **23**(2), 304–310. https://doi.org/10.1093/jamia/ocv080

21. He, K., Zhang, X., Ren, S., Sun, J.: Deep residual learning for image recognition. CoRR, abs/1512.03385 (2015). http://arxiv.org/abs/1512.03385

22. Simonyan, K.: Zisserman, A.: Very deep convolutional networks for large-scale image recognition (2015)

23. Papineni, K., Roukos, S., Ward, T., Zhu, W.: Bleu: a method for automatic evaluation of machine translation. In: Proceedings of the 40th Annual Meeting on Association for Computational Linguistics, pp. 311–318. Association for Computational Linguistics (2012)

24. Lin, C.-Y.: Rouge: a package for automatic evaluation of summaries. In: Proceedings of the ACL-04 Workshop on Text Summarization Branches Out, vol. 8. Barcelona, Spain (2004)

25. Xue, Y., et al.: Multimodal recurrent model with attention for automated radiology report generation. In: Proceedings of the 21st International Conference on Medical Image Computing and Computer Assisted Intervention (MICCAI 2018) (2018)

26. Olah, C.M., Ludwig, A.S.: Feature visualization. Distill (2017)

27. Lipton, Z.C.: The mythos of model interpretability. In: Workshop on Human Interpretability in Machine Learning (WHI 2016) (2016)

Visual Explanation by Unifying Adversarial Generation and Feature Importance Attributions

Martin Charachon[1,2]([✉]), Paul-Henry Cournède[2], Céline Hudelot[2], and Roberto Ardon[1]

[1] Incepto Medical, Paris, France
[2] MICS, Université Paris-Saclay, CentraleSupélec, France

Abstract. Explaining the decisions of deep learning models is critical for their adoption in medical practice. In this work, we propose to unify existing adversarial explanation methods and path-based feature importance attribution approaches. We consider a path between the input image and a generated adversary and associate a weight depending on the model output variations along this path. We validate our attribution methods on two medical classification tasks. We demonstrate significant improvement compared to state-of-the-art methods in both feature importance attribution and localization performance.

Keywords: Explainable AI · Deep learning · Classification · GANs · Medical image

1 Introduction

While deep learning models are nowadays commonly used in the medical domain [1,8,14], a major limitation to their general adoption in daily clinical routines is the lack of explanations with respect to their predictions [16]. In this work, we focus on visual explanation methods [9,21,30]. They provide an additional image where regions of higher importance -for the model prediction- are expected to correlate with pathology location when the model prediction is correct or fail to otherwise. Thus, visual explanation images help clinicians to assign a level of confidence to a model.

Several contributions have been made [4,7,20,23,28] based on the generation of adversaries (images closely related to input images but that have a different model prediction), where visual explanation is defined as the difference between this adversary and the input image, or its regularized version [4,23]. They perform particularly well in highlighting global relevant regions for classification models that match expert expectations e.g. localized pathology. Despite efforts to produce visual explanations only containing relevant information to the model, residual noise still remains. On the other hand, path-based methods [17], which derive visual explanation from pixel-wise derivatives of the model,

© Springer Nature Switzerland AG 2021
M. Reyes et al. (Eds.): iMIMIC 2021/TDA4MedicalData 2021, LNCS 12929, pp. 44–55, 2021.
https://doi.org/10.1007/978-3-030-87444-5_5

are built to detect pixel regions with high impact on the model prediction. They generally result in very noisy outputs. The main contribution of this paper is to unify both adversarial and path-based feature attribution methods (Sect. 3). In the spirit of [26], we go a step further and follow a path between the input image and its adversary. We associate to each element of this path a weight reflecting feature importance through the model gradient. Visual explanation is then defined as the sum of all contributions along the path. We validated our method (Sect. 4) on two classification tasks and publicly available data sets: slice classification for brain tumors localization in MRI, and pneumonia detection on chest X-Rays.

2   Related Works

Within the numerous contributions in visual explanation for classification models [5,9,10,19,21], our work is at the crossroads of two major families, gradient attribution methods and adversarial methods. In the following, let f be the classifier to explain, $\mathbf{x}$ the input image to which the classifier f is applied and $\mathcal{E}$ the visual explanation image.

Gradient attribution methods [21,24–26,29] generate visual explanations by associating an importance value with each pixel of $\mathbf{x}$ using back propagation of the classifier's gradient: $\frac{\partial f}{\partial \mathbf{x}}(\mathbf{x}) = \left\{ \frac{\partial f}{\partial x_i}(\mathbf{x}) \right\}_i$ where index i enumerates all pixels of $\mathbf{x}$. In particular, [26] introduced the idea of integrating this gradient along a path of images to define the explanation as $\mathcal{E} = (\mathbf{x} - \mathbf{z}) \int_0^1 \frac{\partial f(\mathbf{z} + \lambda(\mathbf{x} - \mathbf{z}))}{\partial \mathbf{x}} \, d\lambda$ (where $\mathbf{z}$ is the null image or random noise). These methods generally produce visual explanations that contain relevant regions but are often corrupted by noise and have limited quality (see Table 1).

In a different perspective, adversarial methods [3,4,7,23,28] propose to generate visual explanation by comparing the input image with a "close" generated adversarial example $\mathbf{x_a}$ and defining visual explanation by

$$\mathcal{E}(\mathbf{x}) = |\mathbf{x} - \mathbf{x_a}|. \tag{1}$$

The prediction of classifier f for the adversary $\mathbf{x_a}$ is expected to be different from that of the original image (e.g. $f(\mathbf{x_a}) = 1 - f(\mathbf{x})$ for a binary classifier) and only differ from the input image on regions that are critical for the decision of f. Moreover, [3,23] advocate that $\mathbf{x_a}$ should belong to the distribution of real images (for instance by leveraging domain translation techniques [31]) in order to explain the behavior of f within the distribution of images it is expected to work on. Several works [4,18,23] also use the notion of a "stable" image $\mathbf{x_s}$ defined as the closest element to $\mathbf{x}$ (in the sense of norm $L_{1,2}$) generated by a comparable generation process as for $\mathbf{x_a}$ but classified by f like $\mathbf{x}$ ($f(\mathbf{x}) = f(\mathbf{x_s})$). The goal is to reduce reconstruction errors due to the generation process which are irrelevant to the visual explanation, then defined as $\mathcal{E} = |\mathbf{x_s} - \mathbf{x_a}|$. These methods have high performances in localizing pathological regions when acting on a classifier f trained to detect if there is any pathology (Table 1).

3　Method

Compared to gradient attribution methods, adversarial approaches generate outputs that are smoother and more localized (Table 1). However, no adversarial approach explicitly enforces that visual explanation values translate into importance values for f at the pixel level or at any higher scale. For instance, suppose $\mathbf{x}$ is a CT-scan and classifier f is influenced by regions containing bone tissues (which have high intensity in CT-scans). These regions should then be attenuated in the adversary $\mathbf{x_a}$ and appear with high intensity in the difference $\mathcal{E} = |\mathbf{x} - \mathbf{x_a}|$. This high intensity may not be directly related to the relative importance of bone regions for f but only result from their original intensity in $\mathbf{x}$. This case would poorly perform using the AOPC metric introduced in [17]. Experimentally, even when using $\mathbf{x_s}$, it is sometimes impossible to remove all irrelevant regions for f.

3.1　Combining Gradient Attribution and Adversarial Methods

Consider an image $\mathbf{x}$ or its "stable" generation $\mathbf{x_s}$ (depending on the chosen adversarial method) and its generated adversary $\mathbf{x_a}$. Following [23,26] we consider a differential path γ mapping elements $\lambda \in [0, 1]$ to the space of real images [4,23] and satisfying $\gamma(0) = \mathbf{x}$ and $\gamma(1) = \mathbf{x_a}$. From Eq. (1) we have

$$\mathcal{E}(\mathbf{x}) = |\mathbf{x_a} - \mathbf{x}| = \left| \int_0^1 \frac{d\gamma}{d\lambda}(u)du \right|. \tag{2}$$

To enforce a monotonic relationship between high value regions of $\mathcal{E}$ and high importance regions of f, we propose to introduce weights related to the variations of f along the path integral (2). We define these weights (w) at every $u \in [0, 1]$ based on the variations $\dfrac{d(f \circ \gamma)}{d\lambda}(u) = \dfrac{\partial f}{\partial \mathbf{x}}(\gamma(u))\dfrac{d\gamma}{d\lambda}(u)$. Several strategies are possible for w. The expressions studied in Sect. 3.2 can be summarized using a continuous function of two variables F, setting $w(u) = F\left(\frac{\partial f}{\partial \mathbf{x}}(\gamma(u)), \frac{d\gamma}{d\lambda}(u)\right)$.

We then define the visual explanation map as

$$\mathcal{E}_w(\mathbf{x}) = \left| \int_0^1 w(u)\frac{d\gamma}{d\lambda}(u)du \right| = \left| \int_0^1 F\left(\frac{\partial f}{\partial \mathbf{x}}(\gamma(u)), \frac{d\gamma}{d\lambda}(u)\right)\frac{d\gamma}{d\lambda}(u)du \right| \tag{3}$$

Weights w, as well as path γ and its derivative, are of the same dimension as image $\mathbf{x}$, summation and multiplication are thus done pixel-wise.

3.2　Choice of the Path and Regularization

Ideally path γ should be traced on the manifold of real clinical images (as in [23]). In practice, this constraint induces heavy computation burdens to determine the derivative $\frac{d\gamma}{d\lambda}$ [1]. To tackle this issue we use a similar expression as [26] and define $\gamma(\lambda) = \mathbf{x} + \lambda(\mathbf{x_a} - \mathbf{x})$ so that $\frac{d\gamma}{d\lambda} = (\mathbf{x_a} - \mathbf{x})$. Experimentally, even with this simplification, visual explanation maps integrating feature importance (FI)

$$\mathcal{E}_{FI}^{v1}(x) = (\mathbf{x_a} - \mathbf{x})^2 \left| \int_0^1 \frac{\partial f}{\partial \mathbf{x}}(\gamma(u)) du \right|$$

$$\mathcal{E}_{FI}^{v2}(x) = (\mathbf{x_a} - \mathbf{x})^2 \int_0^1 \left| \frac{\partial f}{\partial \mathbf{x}}(\gamma(u)) \right| du \tag{4}$$

outperform state-of-the-art methods (Sect. 4). $\mathcal{E}_{FI}^{v1}$ is obtained by setting $F:$ $(x, y) \to x.y$ and is used as baseline. To take into account all derivatives regardless of their signs we set $F: (x, y) \to |x|.y$ and obtain $\mathcal{E}_{FI}^{v2}$. Finally, despite the accumulation of gradients along the linear path between $\mathbf{x}$ and $\mathbf{x}_a$, $\mathcal{E}_{FI}^{v1}$ and $\mathcal{E}_{FI}^{v2}$ tend to be noisy. We thus introduce a regularized version

$$\mathcal{E}_{FI,k_\sigma}^{v2}(x) = \int_0^1 \left((\mathbf{x_a} - \mathbf{x})^2 \left| \frac{\partial f}{\partial \mathbf{x}}(\gamma(u)) \right| \right) * k_\sigma du \tag{5}$$

where k_σ is a centered Gaussian kernel of variance σ. In our experiments $\mathcal{E}_{FI,k_\sigma}^{v2}$ is competitive with $\mathcal{E}_{FI}^{v1}$ and $\mathcal{E}_{FI}^{v2}$ for both the AOPC metric and the feature relevance score on feature importance evaluation [13,17] while improving pathology localization performance.

4 Experiments and Results

4.1 Datasets and Models

Slice Classification for Brain Tumor Detection - We use the dataset of Magnetic Resonance Imaging (MRI) for brain tumor segmentation from the Medical Segmentation Decathlon Challenge [22]. Only using the contrasted T1-weighted (T1gd) sequence, we transform the 4-level annotations into binary masks. We then resize the 3D volumes and corresponding binary masks from $155 \times 240 \times 240$ to $145 \times 224 \times 224$. From these resized masks we affect a class label to each single slice of the volume along the axial axis. Class 1 is given to a slice if at least 10 pixels (0.02%) are tumorous. Then, the objective is to train a classifier to detect slices with tumors. As additional prepossessing, we remove all slices outside the body along the axial axis and normalize all slice images in [0, 1]. 46900 slices are used for training, 6184 for validation and 9424 for test with 25% of slices with tumors (in each set).

---

[1] As in [3,23], consider an encoder(E)-generator(G) architecture. E (resp. G) maps from (resp. to) the space of real images ($\subset \mathbb{R}^n$) to (resp. from) an encoding space ($\subset \mathbb{R}^k$). The *real* images path γ can for instance be defined as $\gamma : \lambda \to G(z_\mathbf{x} + \lambda(z_{\mathbf{x_a}} - z_\mathbf{x}))$, where $z_\mathbf{x} = E(\mathbf{x})$ and $z_{\mathbf{x_a}} = E(\mathbf{x_a})$. It follows that $\frac{d\gamma}{d\lambda} = \frac{\partial G}{\partial z}(z_\mathbf{x} + \lambda(z_{\mathbf{x_a}} - z_\mathbf{x}))(z_{\mathbf{x_a}} - z_\mathbf{x})$. But $\frac{\partial G}{\partial z}$ is a vector of dimension $n.k$ which easily reaches a magnitude of 10^9 that is to be computed at several values of λ.

Pneumonia Detection - We also use a chest X-Ray dataset from the available RSNA Pneumonia Detection Challenge which consists of X-Ray dicom exams extracted from the NIH CXR14 dataset [27]. We constitute our binary database with 8851 healthy and 6012 pathological exams.As in [4], we only keep the healthy and pathological exams constituting a binary database of 14863 samples (8851 healthy/6012 pathological). Images are rescaled from 1024×1024 to 224×224 and normalized to $[0, 1]$. Bounding box annotations around opacities are provided for pathological cases. The split consists in 11917 exams in train, 1495 in validation and 1451 in test.

Classifier - For the two problems, we train an adapted ResNet50 [11] using the Adam optimizer [12] with an initial learning rate of 1e–4, and minimizing a weighted binary cross-entropy. We recover the ResNet50 backbone trained on ImageNet [6], and add a dropout layer (rate = 0.3), then two fully connected layers with respectively 128 and 1 filters. We also introduce random geometric transformations such as flip, rotation, zoom or translation during the training. The classifiers respectively achieve 0.975 and 0.974 AUC scores on brain tumor and pneumonia detection problems.

4.2 Attribution Techniques and Implementation Details

Baseline Methods - We compare our method to several baselines and state-of-the-art visual explanation approaches:

(1) Gradient-based (or CAM-based): Gradient [21], Integrated Gradient [26] (IG), GradCAM [19] (GCAM).
(2) Perturbation-based: Mask Perturbation [9] (MPert), Mask Generator [5] (MGen), Similar and Adversarial Generations [4] (SAGen).
(3) Adversary based: the two variations proposed in [3]: CyCE and SyCE. In CyCE, the explanation is computed with the input image $|\mathbf{x} - \mathbf{x_a}|$, while the stable image is used in SyCE version ($|\mathbf{x_s} - \mathbf{x_a}|$).

For MPert [9], we look for a mask of size 56×56, and filter it after upsampling ($\sigma = 3$). We use gaussian blur ($\sigma = 5$) to perturbate the input through the mask. Masks are regularized with total variation and finally obtained after 150 iterations. We use a ResNet-50 backbone pre-trained on the task as the encoder part of MGen, then we basically follow the UNet-like [15] architecture and training proposed in [5]. MGen produces mask of size 112×112 that are upsampled to 224×224. For SAGen [4], we use a UNet-like architecture as the common part of the generators and two separated final convolutional blocks for respectively stable and adversarial generations.

Our Methods - For the different variations $\mathcal{E}_{FI}^{v1}$, $\mathcal{E}_{FI}^{v2}$ and $\mathcal{E}_{FI,k_\sigma}^{v2}$, the integral is approximated using a Riemann sum. For instance, in SyCE, $\mathcal{E}_{FI}^{v2}$ is computed through:

$$\mathcal{E}_{FI}^{v2}(x) \approx \frac{(\mathbf{x_s} - \mathbf{x_a})^2}{M} \sum_{m=1}^{M} \left| \frac{\partial f}{\partial \mathbf{x}}(\gamma_m) \right| \qquad (6)$$

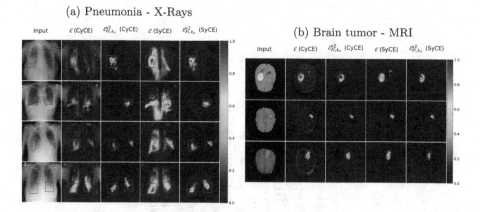

(a) Pneumonia - X-Rays

(b) Brain tumor - MRI

Fig. 1. Attribution maps visualization. For (a) and (b), left columns show the input image with red contours indicating pathology localization. Other columns show visualization explanations generated by our regularized method $\mathcal{E}_{FI,k_\sigma}^{v2}$ and two adversarial approaches of [3]. (Color figure online)

where $\gamma_m = \mathbf{x_s} + \frac{m-1/2}{M}(\mathbf{x_a} - \mathbf{x_s})$, and M is the number of steps in the Riemann sum. In our experiments, we take $M = 50$. For the regularized version, we apply a Gaussian filtering of kernel 28×28 and $\sigma = 2$.

4.3   Results

Pathology Localization. For strong classifiers f, as it is the case here (see Sect. 4.1), we expect the feature attributions to match experts annotations as much as possible. We measure the intersection over union (IoU)), and the proportion of attribution maps that lays outside the ground truth annotations, also called False Negative Rate (FNR). Both metrics are computed between ground truth annotations and thresholded binary explanation maps. As in [3], the attribution maps are thresholded given percentile values. We only display IoU and FNR scores for percentile values that mostly represent the size distribution of ground truth annotations. Table 1 reports the localization scores. First, adversarial generation methods CyCE and especially SyCE outperform common state-of-the-art approaches. For both SyCE and CyCE, our proposed methods $\mathcal{E}_{FI}^{v1}$, $\mathcal{E}_{FI}^{v2}$ and $\mathcal{E}_{FI,k_\sigma}^{v2}$ significantly improve localization performances compared to the baseline $\mathcal{E}$ (shown in blue or red in Table 1), or are at least competitive ($\mathcal{E}_{FI}^{v1}$ for SyCE). In the two adversarial generation approaches, $\mathcal{E}_{FI}^{v2}$ outperforms $\mathcal{E}_{FI}^{v1}$, but the regularized version $\mathcal{E}_{FI,k_\sigma}^{v2}$ is the best localizer (red). Figure 1a and fig1b display qualitative results comparing visual explanation from baseline methods CyCE and SyCE with our regularized approach $\mathcal{E}_{FI,k_\sigma}^{v2}$. It visually supports the localization results shown in Table 1. Our method focuses only on important region for the classifier which also correlate with human annotations, and remove residual errors remaining in baseline attribution maps (especially for CyCE).

Table 1. Localization results. Different attribution methods on Pneumonia detection and Brain tumor problems. IoU (higher is better) and FNR (lower is better) scores are given at representative percentile values for each problem.

	Metric	Perc.	Grad.	IG	GCAM	MPert	MGen	SAGen	CyCE				SyCE			
									ε	ε_{F1}^{v1}	ε_{F1}^{v2}	$\varepsilon_{F1,k_\sigma}$	ε	ε_{F1}^{v1}	ε_{F1}^{v2}	$\varepsilon_{F1,k_\sigma}$
Pneumonia	IoU ↑	90	0.187	0.170	0.195	0.204	0.208	0.232	0.221	0.269	0.300	**0.321**	0.299	0.289	0.323	0.335
		95	0.152	0.136	0.138	0.154	0.169	0.173	0.191	0.232	0.259	**0.278**	0.244	0.242	0.271	0.285
		98	0.097	0.086	0.070	0.087	0.103	0.097	0.116	0.149	0.163	**0.175**	0.151	0.150	0.165	0.177
	FNR ↓	90	0.639	0.698	0.645	0.623	0.620	0.584	0.596	0.532	0.494	**0.471**	0.492	0.507	0.469	0.457
		95	0.584	0.653	0.618	0.576	0.542	0.535	0.494	0.421	0.379	**0.352**	0.399	0.407	0.363	0.344
		98	0.508	0.603	0.593	0.537	0.461	0.495	0.430	0.321	0.285	**0.257**	0.319	0.314	0.275	0.250
Brain Tumor	IoU ↑	98	0.154	0.238	0.173	0.290	0.318	0.330	0.322	0.337	0.376	**0.428**	0.411	0.404	0.426	0.432
		99	0.131	0.196	0.115	0.263	0.274	0.284	0.270	0.288	0.322	**0.363**	0.348	0.338	0.357	0.364
	FNR ↓	98	0.744	0.621	0.715	0.580	0.534	0.525	0.542	0.518	0.481	0.433	0.462	0.462	0.446	**0.441**
		99	0.687	0.536	0.701	0.451	0.413	0.408	0.440	0.401	0.358	0.311	0.344	0.347	0.329	**0.324**

Table 2. Feature Relevance Score R. Comparing the different attributions methods on Pneumonia detection and Brain tumor problems. The score R is given for specific predicted class 0 and 1 as well as for the two combined (ALL).

	Pred. class	Random	Grad.	IG	GCAM	MPert	MGen	SAGen	CyCE $\mathcal{E}$	$\mathcal{E}_{FI}^{v1}$	$\mathcal{E}_{FI}^{v2}$	$\mathcal{E}_{FI,k_\sigma}^{v2}$	SyCE $\mathcal{E}$	$\mathcal{E}_{FI}^{v1}$	$\mathcal{E}_{FI}^{v2}$	$\mathcal{E}_{FI,k_\sigma}^{v2}$
Pneumonia	All	0.118	0.483	0.385	0.473	0.475	0.339	0.463	0.636	0.672	**0.686**	**0.686**	0.593	0.657	0.679	0.712
	0	0.172	0.597	0.562	0.563	0.582	0.304	0.504	0.771	0.830	**0.839**	0.810	0.758	0.823	0.844	0.835
	1	0.049	0.037	0.009	0.363	0.255	0.368	0.084	0.357	0.359	0.376	**0.462**	0.298	0.342	0.358	0.506
Brain tumor	All	0.066	0.563	0.599	0.714	0.631	0.689	0.736	0.638	0.675	0.693	**0.703**	0.754	0.766	0.778	0.788
	0	0.077	0.509	0.476	0.712	0.460	0.580	0.686	0.486	0.535	0.569	**0.570**	0.702	0.714	0.737	0.753
	1	0.055	0.614	0.708	0.715	0.780	0.787	0.783	0.767	0.795	0.801	**0.820**	0.803	0.815	0.817	0.821

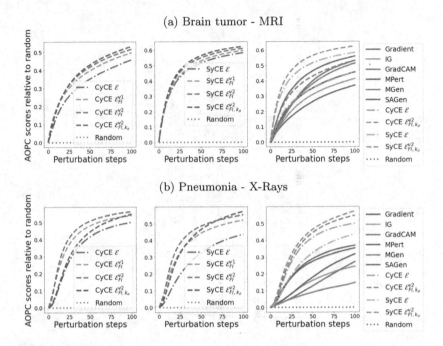

Fig. 2. AOPC scores relative to random baseline. The first two columns: baseline adversarial approaches CyCE $\mathcal{E}$ (left) and SyCE $\mathcal{E}$ (middle) compared with our proposed methods. Last column: comparision of baseline $\mathcal{E}$ and regularized version $\mathcal{E}_{FI,k_\sigma}^{v2}$ with other state-of-the-art methods. Results are given for the two classification problems. The higher area, the better. (Color figure online)

Feature Relevance Evaluation. Although localization performance enables human experts to assess the quality of the visual explanation, it is not enough to translate the importance of features for the classifier. High localization performance does not reflect the capacity of the visual explanation to order regions of the input image w.r.t their importance for the model decision. It only reports on its capacity to find these regions. To evaluate feature importance for the classifier, we use two metrics based on input degradation techniques [17]: **(i)** the area over the perturbation curve (AOPC) by progressively perturbing the input, starting with the most relevant regions of the explanation map first (introduced in [17]); and **(ii)** the feature relevance score (R) proposed in [13] which combines degradation (most relevant first) and preservation (least relevant first) impacts w.r.t. the classifier. For both a perturbation method must be set. In our experiments, we use an adversarial perturbation (as in [2]). Other types of perturbations (replacement by zero, replacement by noise) generate images outside of the training distribution and break down all visual explanation methods, rendering their evaluation impossible. The adversarial perturbation process for these metrics is independent from adversary generations in adversarial-based visual explanations to produce fair evaluation comparisons. It basically follows the image-to-image translation approach proposed in [20]. The two metrics are computed on a balanced subset of 1000 images of the test set.

Table 2 shows the feature relevance score R for specific ("0" or "1") and combined ("ALL") predicted classes. Then, Figs. 2a and b show the evolution of the AOPC score on the two classification problems for the different visual explanation approaches relative to a random baseline. (i) Our proposed methods improve adversarial generation baselines CyCE and SyCE for both relevance score on the two predicted classes (blue and red in Table 2), and the AOPC metric (red, green and yellow curves compared to the blue one on the first two columns of Figs. 2a and b). (ii) The regularized version $\mathcal{E}_{FI,k_\sigma}^{v2}$ (red curve) is competitive with $\mathcal{E}_{FI}^{v1}$ and $\mathcal{E}_{FI}^{v2}$ (or even outperformed them on Brain tumor problem). (iii) Our methods outperform state-of-the-art approaches (last column in the AOPC figures), especially the ones based on SyCE adversarial generation (see Table 2).

5 Conclusion

We propose a unification of adversarial visual explanation methods and path-based feature attribution approaches. Using a linear path between the input image and its generated adversary, we introduce a tractable method to assign a weight along this path translating variations of the classifier output. Our method better assesses feature importance attribution compared to both adversarial generation approaches and path-based feature attribution methods. We also improve relevant regions localization performances by reducing the residual reconstruction errors inherent to adversarial generation methods.

References

1. Bien, N., et al.: Deep-learning-assisted diagnosis for knee magnetic resonance imaging: development and retrospective validation of MRNet. PLoS Med. **15**, 1002699 (2018)
2. Chang, C.H., Creager, E., Goldenberg, A., Duvenaud, D.K.: Explaining image classifiers by counterfactual generation. In: ICLR (2019)
3. Charachon, M., Cournède, P., Hudelot, C., Ardon, R.: Leveraging conditional generative models in a general explanation framework of classifier decisions. In: ArXiv (2021)
4. Charachon, M., Hudelot, C., Cournède, P.H., Ruppli, C., Ardon, R.: Combining similarity and adversarial learning to generate visual explanation: Application to medical image classification. In: ICPR (2020)
5. Dabkowski, P., Gal, Y.: Real time image saliency for black box classifiers. In: NIPS (2017)
6. Deng, J., Dong, W., Socher, R., Li, L.J., Li, K., Fei-Fei, L.: ImageNet: a large-scale hierarchical image database. In: CVPR (2009)
7. Elliott, A., Law, S., Russell, C.: Adversarial perturbations on the perceptual ball(2019). ArXiv arXiv:1912.09405

8. Esteva, A., et al.: Dermatologist-level classification of skin cancer with deep neural networks. Nature **542**, 115–118 (2017)
9. Fong, R.C., Vedaldi, A.: Interpretable explanations of black boxes by meaningful perturbation. In: ICCV (2017)
10. Goyal, Y., Wu, Z., Ernst, J., Batra, D., Parikh, D., Lee, S.: Counterfactual visual explanations. In: ICML (2019)
11. He, K., Zhang, X., Ren, S., Sun, J.: Identity mappings in deep residual networks. In: ECCV (2016)
12. Kingma, D.P., Ba, J.: Adam: a method for stochastic optimization. In: ICLR (2015)
13. Lim, D., Lee, H., Kim, S.: Building reliable explanations of unreliable neural networks: locally smoothing perspective of model interpretation. In: CVPR (2021)
14. Pratt, H., Coenen, F., Broadbent, D.M., Harding, S.P., Zheng, Y.: Convolutional neural networks for diabetic retinopathy. Procedia Comput. Sci. **90**, 200–205 (2016). https://doi.org/10.1016/j.procs.2016.07.014, https://www.sciencedirect.com/science/article/pii/S1877050916311929, 20th Conference on Medical Image Understanding and Analysis (MIUA 2016)
15. Ronneberger, O., Fischer, P., Brox, T.: U-net: convolutional networks for biomedical image segmentation. In: MICCAI (2015)
16. Rudin, C.: Stop explaining black box machine learning models for high stakes decisions and use interpretable models instead. Nat, Mach. Intell. **1**(5), 206–215 (2019)
17. Samek, W., Binder, A., Montavon, G., Lapuschkin, S., Müller, K.: Evaluating the visualization of what a deep neural network has learned. In: IEEE Transactions on Neural Networks and Learning Systems (2017)
18. Seah, J.C.Y., Tang, h.J.S.N., Kitchen, A., Gaillard, F., Dixon, A.F.: Chest radiographs in congestive heart failure: visualizing neural network learning. Radiology **290**, 514-522 (2019)
19. Selvaraju, R.R., Cogswell, M., Das, A., Vedantam, R., Parikh, D., Batra, D.: Grad-CAM: visual explanations from deep networks via gradient-based localization. In: ICCV (2017)
20. Siddiquee, M.R., et al.: Learning fixed points in generative adversarial networks: from image-to-image translation to disease detection and localization. In: ICCV, pp. 191–200 (2019)
21. Simonyan, K., Vedaldi, A., Zisserman, A.: Deep inside convolutional networks: visualising image classification models and saliency maps. In: ICLR (2014)
22. Simpson, A., et al.: A large annotated medical image dataset for the development and evaluation of segmentation algorithms. ArXiv arXiv:1902.09063 (2019)
23. Singla, S., Pollack, B., Chen, J., Batmanghelich, K.: Explanation by progressive exaggeration. In: ICLR (2020)
24. Smilkov, D., Thorat, N., Kim, B., Viégas, F.B., Wattenberg, M.: Smoothgrad: removing noise by adding noise. ArXiv arXiv:1706.03825 (2017)
25. Springenberg, J.T., Dosovitskiy, A., Brox, T., Riedmiller, M.A.: Striving for simplicity: the all convolutional net. In: ICLR (2015). arXiv:1412.6806
26. Sundararajan, M., Taly, A., Yan, Q.: Axiomatic attribution for deep networks. In: ICML (2017)
27. Wang, X., Peng, Y., Lu, L., Lu, Z., Bagheri, M., Summers, R.M.: Chestx-ray8: Hospital-scale chest x-ray database and benchmarks on weakly-supervised classification and localization of common thorax diseases. In: CVPR (2017)
28. Woods, W., Chen, J., Teuscher, C.: Adversarial explanations for understanding image classification decisions and improved neural network robustness. Nat. Mach. Intell. **1** (2019)

29. Xu, S.Z., Venugopalan, S., Sundararajan, M.: Attribution in scale and space. In: CVPR (2020)
30. Zhou, B., Khosla, A., Lapedriza, À., Oliva, A., Torralba, A.: Learning deep features for discriminative localization. In: CVPR (2016)
31. Zhu, J.Y., Park, T., Isola, P., Efros, A.A.: Unpaired image-to-image translation using cycle-consistent adversarial networks. In: ICCV (2017)

The Effect of the Loss on Generalization: Empirical Study on Synthetic Lung Nodule Data

Vasileios Baltatzis[1,2(✉)], Loïc Le Folgoc[2], Sam Ellis[1],
Octavio E. Martinez Manzanera[1], Kyriaki-Margarita Bintsi[2], Arjun Nair[3],
Sujal Desai[4], Ben Glocker[2], and Julia A. Schnabel[1,5,6]

[1] School of Biomedical Engineering and Imaging Sciences, King's College London,
London, UK
vasileios.baltatzis@kcl.ac.uk
[2] BioMedIA, Department of Computing, Imperial College London, London, UK
[3] Department of Radiology, University College London, London, UK
[4] The Royal Brompton and Harefield NHS Foundation Trust, London, UK
[5] Technical University of Munich, Munich, Germany
[6] Helmholtz Center Munich, Munich, Germany

Abstract. Convolutional Neural Networks (CNNs) are widely used for image classification in a variety of fields, including medical imaging. While most studies deploy cross-entropy as the loss function in such tasks, a growing number of approaches have turned to a family of contrastive learning-based losses. Even though performance metrics such as accuracy, sensitivity and specificity are regularly used for the evaluation of CNN classifiers, the features that these classifiers actually learn are rarely identified and their effect on the classification performance on out-of-distribution test samples is insufficiently explored. In this paper, motivated by the real-world task of lung nodule classification, we investigate the features that a CNN learns when trained and tested on different distributions of a synthetic dataset with controlled modes of variation. We show that different loss functions lead to different features being learned and consequently affect the generalization ability of the classifier on unseen data. This study provides some important insights into the design of deep learning solutions for medical imaging tasks.

Keywords: Distribution shift · Interpretability · Contrastive learning

1 Introduction

Deep learning methods and particularly Convolutional Neural Networks (CNNs) are, currently, the backbone of most state-of-the-art approaches for medical image classification tasks. The performance of machine learning techniques, however, can drop significantly when the test data are from a different distribution than the training data, which is common in real-world applications, such as medical images originating from different hospitals, acquired with different protocols, or when there is a lack of or variation in high quality annotations.

M. Reyes et al. (Eds.): iMIMIC 2021/TDA4MedicalData 2021, LNCS 12929, pp. 56–64, 2021.
https://doi.org/10.1007/978-3-030-87444-5_6

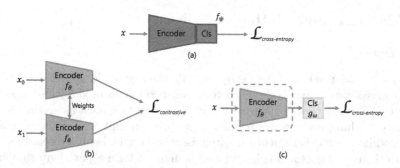

Fig. 1. Overview of the used CNN architectures, which are based on LeNet-5 [4]. (a) An encoder-classifier network trained end-to-end with CE. (b) A siamese network trained with a contrastive loss. (c) The encoder from (b) is frozen and a classifier (Cls), identical to the one used in (a), is trained on top of it with CE.

Motivated by these obstacles, we study the effect of data variation utilizing a synthetic dataset with specific modes of variation. The limitations that arise from a synthetic dataset are clear since its simplified nature does not reflect the complexity of a real, clinical dataset. However, it is exactly this complexity that we are trying to avoid, as it would not allow us to evaluate very specific scenarios in terms of controlling the exact characteristics of the training and test distributions. This fully controlled setting allows us to create training and test distributions with similar or contrasting characteristics. We leverage this dataset to explore the subtlety of the differences between training and test distributions that is sufficient to hamper performance. We do not suggest that this simplification can lead to a direct application on disease classification but rather our primary intent is to investigate the behavior of CNNs under certain distribution shifts at test time to a very fine level of detail, which would be impossible to achieve if we shifted to a real-world medical imaging dataset. To examine thoroughly the features learned by a CNN and how these can influence the performance for out-of-distribution (OOD) test samples, we utilize principal component analysis (PCA) and saliency maps. Additionally, we study the increasingly popular contrastive learning-based losses [2] proposed in recent work [1,8]. Here, we investigate the differences between a cross-entropy (CE) loss and a contrastive loss, in terms of both performance and resulting CNN features.

Our contributions can be summarized as follows: 1) We design a synthetic dataset with two modes of variation (binary shape class and average intensity of the shape appearance) inspired by the real world application of lung nodule classification; 2) We conduct an experimental study to explore the effects of two different loss functions (CE and contrastive) on the learned CNN features (under different training distributions) and the impact on OOD generalization; 3) We use a variety of performance metrics (accuracy, sensitivity, specificity) and visualizations (PCA, saliency maps) to support and evaluate our findings. Our findings and insights will be of interest to practitioners designing machine learning solutions for medical imaging applications.

2   Materials and Methods

2.1   Data

The synthetic dataset used here is inspired by the real world application of lung nodule classification and is designed based on two modes of variation. The first mode is the shape class, which is binary. Abnormalities, such as spikes, on the perimeter of lung nodules are termed as spiculation and often indicate malignancy, while a smoother outline is often associated with benign disease [5]. We refer to the two classes as malignant and benign, to form a paradigm similar to lung nodules. The second mode of variation is appearance represented by the average intensity of the pixels within each shape. The values range from 110 to 200 with noise added in 10-point increments, thus giving 10 possible values for this mode, while the background intensities remain fixed for all samples. The synthetic data have been constructed by manually drawing two base shapes (benign vs malignant) from which the experimental dataset is generated using random spatial transformations produced by a combined affine and non-rigid FFD-based transformation model. With $\mathcal{D}_{mode}(i_{mal}, i_{ben})$, we denote a distribution where the average foreground intensity of the malignant and benign shapes is i_{mal} and i_{ben} respectively, while $mode$ refers to either the training or the test set.

2.2   Neural Network Architectures and Loss Functions

We consider two different losses, a CE loss and a contrastive loss, and consequently two neural network architectures that facilitate the two losses (Fig. 1). For simplicity we consider a binary classification task. Both architectures are based on the well-established LeNet-5 [4]. For the first approach, we use a combined encoder-classifier network f_ψ with parameters ψ. It is trained end-to-end, given input image X and label y, via the CE loss (Eq. (1)):

$$\mathcal{L}_{CE} = -y\ log(f_\psi(X)) - (1-y)log(1 - f_\psi(X)) \tag{1}$$

For the second approach we use a Siamese network as in [2], trained in two stages. In the first stage, the network is composed of two copies of the encoder f_θ that share the same weights θ. The input for this system is a pair of images (X_0, X_1) with labels (y_0, y_1) that go through the encoders to produce the representations $f_\theta(X_0)$ and $f_\theta(X_1)$, which are then fed into the contrastive loss defined in Eq. (2):

$$\mathcal{L}_{contr} = \begin{cases} d_\theta(X_0, X_1)^2, & \text{if } y_0 = y_1 \\ \{max(0, m - d_\theta(X_0, X_1))\}^2, & \text{if } y_0 \neq y_1 \end{cases} \tag{2}$$

$$\text{where } d_\theta(X_0, X_1) = \|f_\theta(X_0) - f_\theta(X_1)\|_2. \tag{3}$$

Table 1. Quantitative results for the three experimental scenarios. We report accuracy (Acc), sensitivity (SE) and specificity (SP), for each of the training distribution ($\mathcal{D}_{tr}$), test distribution ($\mathcal{D}_{te}$) and loss combinations that are described in Sect. 3. The results that appear on the table correspond to the performance on ($\mathcal{D}_{te}$). For the training set, all metrics have a value of 1.00 and therefore are not reported on the table.

	$\mathcal{D}_{tr}$	$\mathcal{D}_{te}$	Loss	Acc	SE	SP
Experimental scenario 1	150,150	130,170	CE	1.00	1.00	1.00
	150,150	170,130	CE	0.62	0.87	0.37
	150,150	130,170	Contrast	1.00	1.00	1.00
	150,150	170,130	Contrast	0.15	0.30	0.00
Experimental scenario 2	180,160	150,190	CE	0.94	0.90	0.98
	180,160	190,150	CE	0.96	0.94	0.98
	180,160	150,190	Contrast	0.27	0.01	0.53
	180,160	190,150	Contrast	1.00	1.00	1.00
Experimental scenario 3	180,150	150,190	CE	0.59	0.35	0.83

The loss function minimizes the representation-space distance of Eq. (3) between samples of the same class, while maximizing (bounded by the margin m) the distance between samples of different classes. In the second stage, the encoder f_θ is frozen. We then add a classifier g_ω, with parameters ω, that uses the representations $f_\theta(X)$ as input to perform the classification task. Similarly to the first approach, the encoder f_θ uses an image X and a label y as input and the classifier g_ω is trained with the CE loss. This way, the contrastive loss is used to pre-train the encoder of the network, thus leading to a different set of features that is used for the classification task, compared to the first approach where training is end-to-end.

3 Experiments and Results

We devise three experimental scenarios to demonstrate the OOD test performance by controlling different aspects of the training distribution. For quantitative evaluation, we use accuracy, sensitivity and specificity. We only report these metrics for the OOD test sets, since at train time they are all 1.00. For qualitative evaluation, we utilize PCA to get a two-dimensional projection of the last layer of the CNN before the classification layer and explore the learned feature space. We also use gradient saliency maps [7] to investigate the areas of the input image that contribute most to the CNN prediction.

Training Details. We draw 200 samples from the training distribution, 85% of which are for training and 15% for validation, and another 200 samples from the test distribution for testing. The networks are trained using the Adam optimizer

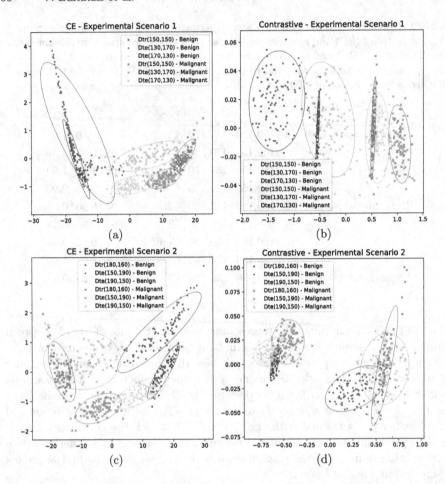

Fig. 2. Experimental Scenario 1 (CE (a), Contrastive (b)) and 2 (CE (c), Contrastive (d)): The PCA projections of the last layer of the CNN before the classification layer. Red hues/'x' denote malignant samples while blue hues/'o' benign. Dark hues are used for the training set $\mathcal{D}_{tr}(i_{mal}, i_{ben})$ and light hues for the two test sets $\mathcal{D}_{te}(i_{mal}, i_{ben})$. The ellipsoids mark two standard deviations distance from the mean of each distribution. In (a) the benign samples of $\mathcal{D}_{te}(170, 130)$ are closer to the malignant cluster of $\mathcal{D}_{tr}(150, 150)$ leading to low specificity. The same is happening in (b) but the malignant samples are also closer to the benign cluster and hence overall performance is low. In (c) there is good generalization in both OOD test sets. In (d) both benign and malignant samples of $\mathcal{D}_{te}(150, 190)$ are close to the opposite cluster of $\mathcal{D}_{tr}(180, 160)$ leading to low performance.

[3] (learning rate $= 10^{-4}$) for 100 epochs and a batch size of 32 samples. The positive and negative pairs for the contrastive loss are dynamically formed within each batch. The margin is chosen to be $m = 1$ based on validation performance, and the Euclidean distance is used as the distance metric. All experiments were conducted using PyTorch [6] and the models were trained on a Titan Xp GPU.

Experimental Scenario 1
Initially, we consider the case where malignant and benign shapes have the same average intensity ($i_{mal} = i_{ben}$). Specifically, we select $\mathcal{D}_{tr}(150, 150)$, since 150 is an intensity in the middle of the distribution of the available intensities, and we use $\mathcal{D}_{te}(130, 170)$ and $\mathcal{D}_{te}(170, 130)$ as these intensities have equal distance from the training distribution for both malignant and benign shapes. Performance metrics can be found in the top four rows of Table 1; PCA projections and saliency maps in Figs. 2a, b and 3a, b, respectively. The CNN fails to classify the OOD test correctly when $i_{mal} > i_{ben}$ for either loss.

Experimental Scenario 2
Next, we consider the case where the average intensities of the whole image (i.e. including the background and not just the pixels inside the shape) are equal for benign and malignant samples ($i_{global\_mal} = i_{global\_ben}$). This happens for $\mathcal{D}_{tr}(180, 160)$, where the average whole image intensity for both malignant and benign images is 117. Equivalently to the first scenario, the OOD test sets come from $\mathcal{D}_{te}(150, 190)$ and $\mathcal{D}_{te}(190, 150)$. The CE trained CNN is able to generalize on both OOD test datasets, while the contrastive loss trained CNN fails when the relationship between i_{mal} and i_{ben} is opposite to what it was in the training distribution. The quantitative results are reported in rows 5–8 of Table 1, while the qualitative results are visualized in Fig. 2c, d (PCA) and c, d (saliency).

Experimental Scenario 3
With the final experiment we want to focus just on one single finding which is the effect of the smallest possible change to the training distribution of the previous scenario (i.e. $\mathcal{D}_{tr}(180, 150)$ instead of $\mathcal{D}_{tr}(180, 160)$), while retaining the same test distributions. For simplicity, we do not focus on analyzing the behaviour of the CNN feature space through saliency maps and PCA projections nor do we use the contrastive loss. We just show results for the CE loss to make sure that we highlight the drop in performance from 0.94 to 0.59 (last row of Table 1) even with the smallest of changes.

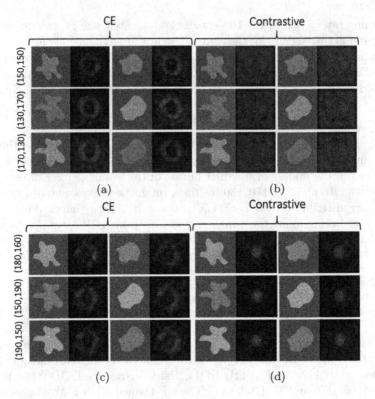

Fig. 3. Experimental Scenario 1 (CE (a), Contrastive (b)) and 2 (CE (c), Contrastive (d)): Example images along with their saliency maps. The first row corresponds to the training set, while the other two to the OOD test sets. In (a) and (b) the activations are spread out across the whole image, even though in (a) some patterns regarding the shape are being picked up. In (c) the activations are strong around the border of the shape, while the background activations are a bit lower compared to (a). In (d) the activations are at their highest at the centre of the image (i.e. within the shape).

4   Discussion

There are three underlying features in the synthetic data distribution that a CNN can try to capture. These are the average intensity of the whole image, the average intensity of the foreground pixels and the shape of the object. From the results of Experimental Scenario 1, where the foreground intensities are equal at train time, we observe that for both losses the CNN fails when the malignant intensity is higher than the benign intensity at test time. This is happening because in this setting, the whole image average intensity is lower for malignant (110) than benign (114) samples, due to the more convex shape of the benign samples, which allows for fewer background pixels. Consequently, the CNN can easily pick up on that feature to distinguish the two classes regardless of the loss function. This can be also confirmed by the saliency maps (Fig. 3a,b), where the activations are spread throughout the whole image, especially for the contrastive loss. The CE loss appears to pick up some patterns in the border of the shape, but the separation of the PCA projections between the two classes is no longer clear for $\mathcal{D}_{te}(170, 130)$ (Fig. 2a).

In Experimental Scenario 2, we remove this discrepancy in the global intensities, and therefore the CNN can no longer use that as a discriminatory feature. In that case, the CNN that was trained with CE is able to generalize in both OOD test sets, which can be confirmed by the PCA projections as well, as they retain the same spatial location as in the training set (Fig. 2c). Hence, it must be capturing the shape information itself. On the other hand, the CNN trained with the contrastive loss learns to distinguish samples based on the average intensity of the pixels of the shape itself, which is evident from the saliency maps, where the most important pixels are the ones in the center of the image (i.e. within the shape) (Fig. 3d). Therefore, the CNN fails when $i_{mal} < i_{ben}$ at test time, since it was $i_{mal} > i_{ben}$ at train time, and the PCA projections for $\mathcal{D}_{te}(150, 190))$ have the opposite mapping to the one for either $\mathcal{D}_{tr}(180, 160))$ or $\mathcal{D}_{te}(190, 150))$.

Finally, in Experimental Scenario 3, we demonstrate that even the slightest change (i.e. reduce i_{ben} to 150 from 160) can have a dramatic impact on the performance of the model on OOD test data, as the accuracy drops from 0.93 to 0.59 for $\mathcal{D}_{te}(150, 190)$. These results indicate how unreliable CNNs can be even when tested on data that are not that far from the training distribution. We demonstrate this failure on a relatively simple dataset. In real applications the relationship between features and the task at hand can be expected to be more complex leading to even worse OOD generalization.

5 Conclusion

Motivated by the important clinical application of lung nodule classification, we have designed a synthetic dataset from a controlled set of variation modes and conducted an exploratory analysis to obtain insights into the learned feature space when trained on different parts of the dataset distribution and how this affects the OOD generalization. The findings indicate that CNN predictions are initially based on the whole image average intensity. When this effect is prohibited, the CNN trained with CE focuses on shape, while the contrastive loss leads the CNN to pick up the average intensity of foreground pixels. Moving forward, we will explore how to constrain the feature space in an automated manner by incorporating application-specific prior knowledge and apply this approach on clinical data.

Acknowledgments. This work is funded by the King's College London & Imperial College London EPSRC Centre for Doctoral Training in Medical Imaging (EP/L015226/1), EPSRC grant EP/023509/1, the Wellcome/EPSRC Centre for Medical Engineering (WT 203148/Z/16/Z), and the UKRI London Medical Imaging & Artificial Intelligence Centre for Value Based Healthcare. The Titan Xp GPU was donated by the NVIDIA Corporation.

References

1. Dou, Q., Castro, D.C., Kamnitsas, K., Glocker, B.: Domain generalization via model-agnostic learning of semantic features. Adv Neural Inf. Process. Syst. **32** (2019), https://github.com/biomedia-mira/masf. arXiv:1910.13580
2. Hadsell, R., Chopra, S., LeCun, Y.: Dimensionality reduction by learning an invariant mapping. In: Proceedings of the IEEE Computer Society Conference on Computer Vision and Pattern Recognition. vol. 2, pp. 1735–1742 (2006). https://doi.org/10.1109/CVPR.2006.100, https://ieeexplore.ieee.org/document/1640964
3. Kingma, D.P., Ba, J.L.: Adam: a method for stochastic optimization. In: 3rd International Conference on Learning Representations (ICLR 2015) - Conference Track Proceedings (2015)
4. LeCun, Y., Bottou, L., Bengio, Y., Haffner, P.: Gradient-based learning applied to document recognition. Proc. IEEE **86**(11), 2278–2323 (1998). https://doi.org/10.1109/5.726791
5. McWilliams, A., Tammemagi, M.C., Mayo, J.R., et al.: Probability of cancer in pulmonary nodules detected on first screening CT. New Engl. J. Med. **369**(10), 910–919 (2013). https://doi.org/10.1056/NEJMoa1214726, https://doi.org/10.1056/NEJMoa1214726
6. Paszke, A., Gross, S., Massa, F., et al.: PyTorch: an imperative style, high-performance deep learning library. In: Advances in Neural Information Processing Systems, vol. 32 (2019)
7. Simonyan, K., Vedaldi, A., Zisserman, A.: Deep inside convolutional networks: visualising image classification models and saliency maps. In: 2nd International Conference on Learning Representations (ICLR 2014) - Workshop Track Proceedings (2014). http://code.google.com/p/cuda-convnet/
8. Winkens, J., Bunel, R., Roy, A.G., et al.: Contrastive training for improved out-of-distribution detection. arXiv preprint 2007.05566 (2020), arXiv:2007.05566

Voxel-Level Importance Maps
for Interpretable Brain Age Estimation

Kyriaki-Margarita Bintsi[1]([✉]), Vasileios Baltatzis[1,2], Alexander Hammers[2],
and Daniel Rueckert[1,3]

[1] BioMedIA, Department of Computing, Imperial College London, London, UK
m.bintsi19@imperial.ac.uk
[2] Biomedical Engineering and Imaging Sciences, King's College London, London, UK
[3] Technical University of Munich, Munich, Germany

Abstract. Brain aging, and more specifically the difference between
the chronological and the biological age of a person, may be a promising
biomarker for identifying neurodegenerative diseases. For this purpose
accurate prediction is important but the localisation of the areas that
play a significant role in the prediction is also crucial, in order to gain
clinicians' trust and reassurance about the performance of a prediction
model. Most interpretability methods are focused on classification tasks
and cannot be directly transferred to regression tasks. In this study,
we focus on the task of brain age regression from 3D brain Magnetic
Resonance (MR) images using a Convolutional Neural Network, termed
prediction model. We interpret its predictions by extracting importance
maps, which discover the parts of the brain that are the most important
for brain age. In order to do so, we assume that voxels that are not
useful for the regression are resilient to noise addition. We implement a
noise model which aims to add as much noise as possible to the input
without harming the performance of the prediction model. We average
the importance maps of the subjects and end up with a population-
based importance map, which displays the regions of the brain that are
influential for the task. We test our method on 13,750 3D brain MR
images from the UK Biobank, and our findings are consistent with the
existing neuropathology literature, highlighting that the hippocampus
and the ventricles are the most relevant regions for brain aging.

Keywords: Brain age regression · Interpretability · Deep learning ·
MR images

1 Introduction

Alzheimer's disease (AD) is the most common cause of dementia [3]. AD leads
to the atrophy of the brain more quickly than healthy aging and is a progressive
neurodegenerative disease, meaning that more and more parts of the brain are
damaged over time. The atrophy primarily appears in brain regions such as
hippocampus, and it afterwards progresses to the cerebral neocortex. At the

© Springer Nature Switzerland AG 2021
M. Reyes et al. (Eds.): iMIMIC 2021/TDA4MedicalData 2021, LNCS 12929, pp. 65–74, 2021.
https://doi.org/10.1007/978-3-030-87444-5_7

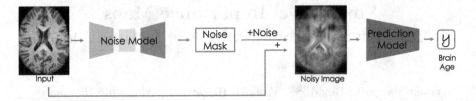

Fig. 1. Overview of the proposed idea. A noise model is trained with the purpose of adding as much noise as possible to the input. The output is a noise mask, in which noise sampled from the standard normal distribution is added. The result is then added to the input image and is used as an input to a pretrained prediction model with frozen parameters. The aim is to create a noisy image that will maximize the noise level while also not harming the performance of the prediction model.

same time, the ventricles of the brain as well as cisterns, which are outside of the brain, enlarge [23].

Healthy aging also results in changing of the brain, following specific patterns [1]. Therefore, a possible biomarker used in AD is the estimation of the brain (biological) age of a subject which can then be compared with the subject's real (chronological) age [8]. People at increased risk can be identified by the deviation between these two ages and early computer-aided detection of possible patients with neurodegenerative disorders can be accomplished. For this reason, a large body of research has focused on estimating brain age, especially using Magnetic Resonance (MR) images, which have long been used successfully in the measurement of brain changes related to age [10].

Recently, deep learning models have proved to be successful on the task of brain age estimation, providing relatively high accuracy. They are designed to find correlation and patterns in the input data and, in the supervised learning setting, associate that with a label, which in our case is the age of the subject. The models are trained on a dataset of MR images of healthy brains, estimating the expected chronological age of the subject. During training, the difference of the chronological age and the predicted age needs to be as small as possible. During test time, an estimated brain age larger than the subject's chronological age indicates accelerated aging, thus pointing to a possible AD patient.

Convolutional Neural Networks (CNNs) are used with the purpose of an accurate brain age estimation while using the minimum amount of domain information since they can process raw image data with little to no preprocessing required. Many studies provide very accurate results, with mean absolute error (MAE) as low as around 2.2 years [16,17,19]. However, most of these approaches purely focus on minimization of the prediction error while considering the network as a black box. Recent studies have started to try to identify which regions are most informative for age prediction. For example, in [6] the authors provided an age prediction for every patch of the brain instead of whole brain. Although the predictions and results presented in [6] were promising, and the localisation was meaningful, the use of large patches meant that the localisation was not very precise. Similar approaches has been explored, such as [4] in which slices of the brain were used instead of patches. In [20], the authors provided an age estimation per voxel but the accuracy of the voxel-wise estimations dropped significantly. In [15], an ensemble of networks and gradient-based techniques [26] were used in order to provide population-based explanation maps. Finally, 3D explanation maps were provided in [5], but the authors used image translation techniques and generative models, such as VAE-GAN networks.

In computer vision, a common approach to investigate black-box models such as CNNs is to use saliency maps, which show which regions the network focuses on, in order to make the prediction. An overview of the existing techniques for explainability of deep learning models in medical image analysis may be found in [25]. Gradient-based techniques, such as guided backpropagation [27], and Grad-CAM [24], make their conclusions based on the gradients of the networks with respect to a given input. Grad-CAM is one of the most extensively used approach and usually results in meaningful but very coarse saliency maps. Gradient-based techniques are focused on classification and to our knowledge they do not work as expected for regression task because they detect the features that contributed mostly to the prediction of a specific class. Occlusion-based techniques [28] have also been widely explored in the literature and they can be used both for classification and regression tasks. The idea behind occlusion techniques is very simple: The performance of the network is explored after hiding different patches of the images, with the purpose of finding the ones that affect the performance the most. It is a promising and straightforward approach but bigger patches provide coarse results. On the other hand, the smaller the patches, the greater the computational and time constraints are, which can be a burden in their application.

A recent approach which leverages the advantages of occlusion techniques while also keeping computational and time costs relatively low is U-noise [14] which uses the addition of noise in the images pixel-wise, while keeping the performance of the network unchanged with the purpose of understanding where the deep learning models focus in order to do their predictions. The authors created interpretability maps for the task of pancreas segmentation using as input 2D images using noise image occlusion. In this paper, inspired by the work described above [14] and the idea that when a voxel is not important for the task, then the addition of noise on this specific voxel will not affect the performance of our network, we make the following contributions: 1) We adapt the architecture of U-noise (Fig. 1), which was originally used for pancreas segmentation, for the task of brain age regression and visualise the parts of the brain that played the most important role for the prediction by training a noise model that dynamically adds noise to the image voxel-wise, providing localised and fine-grained importance maps. 2) We extend the U-noise architecture to 3D instead of 2D to accommodate training with three-dimensional volumetric MR images; 3) We propose the use of an autoencoder-based pretraining step on a reconstruction task to facilitate faster convergence of the noise model; 4) We provide a population-based importance map which is generated by aggregating subject-specific importance maps and highlights the regions of the brain that are important for healthy brain aging.

2   Materials and Methods

2.1   Dataset and Preprocessing

We use the UK Biobank (UKBB) dataset [2] for estimating brain age and extracting the importance maps. UKBB contains a broad collection of brain MR images, such as T1-weighted, and T2-weighted. The dataset we use in this work includes T1-weighted brain MR images from around 15,000 subjects. The images are provided by UKBB skull-striped and non-linearly registered to the MNI152 space. After removing the subjects lacking the information of age, we end up with 13,750 subjects with ages ranging from 44 to 73 years old, 52,3% of whom are females and 47.7% are males. The brain MR images have a size of $182 \times 218 \times 182$ but we resize the 3D images to $140 \times 176 \times 140$ to remove a large part of the background and at the same time address the memory limitations that arise from the use of 3D data, and normalise them to zero mean and unit variance.

2.2   Brain Age Estimation

We firstly train a CNN, the prediction model, f_θ, with parameters θ, for the task of brain age regression. We use the 3D brain MR images as input to 3D ResNet-18, similar to the one used in [6], which uses 3D convolutional layers instead of 2D [11]. The network is trained with a Mean Squared Error (MSE) loss and its output is a scalar representing the predicted age of the subject in years.

2.3   Localisation

An overview of the proposed idea is shown in Fig. 1.

U-Net Pretraining. The prediction model and the noise model have identical architectures (2D U-Net [21]) in [14], as they are both used for image-to-image tasks, which are segmentation and noise mask generation, respectively. Therefore, in that case, the noise model is initialized with the weights of the pretrained prediction model. However, in our case, the main prediction task is not an image-to-image task but rather a regression task and thus, the prediction model's (a 3D ResNet as described above) weights cannot be used to initialize the noise model. Instead, we propose to initially use the noise model f_ψ with parameters ψ, which has the architecture of a 3D U-Net as a reconstruction model, for the task of brain image reconstruction. By pretraining the noise model with a reconstruction task before using it for the importance map extraction task, we facilitate and accelerate the training of our U-noise model, since the network has already learned features relating to the structure of the brain.

The reconstruction model consists of an encoder and a decoder part. It uses as input the 3D MR volumes and its task is to reconstruct the volumes as well as possible. It is trained with a voxel-wise MSE loss. The number of blocks used, meaning the number of downsample and upsample layers, is 5, while the number of output channels after the first layer is 16.

Brain Age Importance Map Extraction. For the extraction of the importance maps, we extend the U-noise architecture [14] to 3D. The idea behind U-noise is that if one voxel is important for the prediction task, in our case brain age regression, the addition of noise on it will harm the performance of the prediction model. On the other hand, if a voxel is not relevant for the task, the addition of noise will not affect the performance.

More specifically, after the pretraining phase with the reconstruction loss, the 3D U-Net, i.e. noise model, is used for the task of extraction of a 3D noise mask that provides a noise level for every voxel of the input 3D image. A sigmoid function is applied so the values of the mask are between 0 and 1. The values are scaled to $[v_{min}, v_{max}]$, where v_{min}, v_{max} are hyperparameters. The rescaled mask is then multiplied by $\epsilon \sim \mathcal{N}(0, 1)$, which is sampled from the standard normal distribution and the output is added to the input image element-wise in order to extract the noisy image. Given an image X, its noisy version can be given by Eq. (1):

$$X_{noisy} = X + f_\psi(X)(v_{max} - v_{min})\epsilon + v_{min}, \tag{1}$$

We then use the noisy image as input for the already trained prediction model with frozen weights, and see how it affects its performance. The purpose is to maximise the noise level in our mask, while simultaneously keeping the performance of our prediction model as high as possible. In order to achieve that, we use a loss function with two terms, the noise term, which is given by $-log(f_\psi(X))$ and motivates the addition of noise for every voxel, and the prediction term, which is an MSE loss whose purpose is to keep the prediction model unchanged. The two loss terms are combined with a weighted sum, which is regulated by the ratio hyperparameter r. It is important to note that at this stage the parameters, θ, of the prediction model, f_θ, are frozen and it is not being trained. Instead, the loss function $\mathcal{L}$, is driving the training of the noise model f_ψ. Given an input image X and label y, the loss function $\mathcal{L}$ takes the form shown in Eq. (2)

$$\mathcal{L} = \underbrace{(f_\theta(X) - y)^2}_{\text{prediction term}} - r\underbrace{log(f_\psi(X))}_{\text{noise term}} \tag{2}$$

The values v_{min}, v_{max}, as well as r are hyperparameters and are decided based on the performance of our noise model on the validation set.

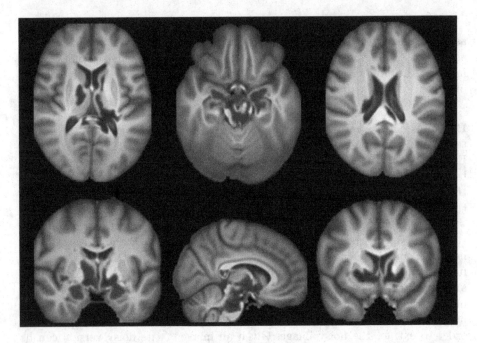

Fig. 2. Six different slices of the population-based importance map on top of the normalised average of the brain MR Images of the test set. The most important parts of the brain for the model's prediction are highlighted in red. The most significant regions for the task of brain aging are mesial temporal structures including the hippocampus, brainstem, periventricular and central areas. The results are in agreement with the relevant literature and previous studies. (Color figure online)

3   Results

From the 13,750 3D brain images, 75% are used for the training set, 10% for the validation set and 15% for the test set. All the networks are trained with backpropagation [22] and adaptive moment estimation (Adam) optimizer [13] with initial learning rate lr = 0.0001, reduced by a factor of 10 every 10 epochs. The experiments are implemented on an NVIDIA Titan RTX using the Pytorch deep-learning library [18].

3.1   Age Estimation

We train the prediction model for 40 epochs using a batch size of 8 using backpropagation. We use MSE between the chronological and biological age of the subject as a loss function. The model achieves a mean absolute error (MAE) of about 2.4 years on the test set.

3.2   Population-Based Importance Maps

The noise model is trained for 50 epochs and with a batch size of 2, on four GPUs in parallel. Different values were tested for hyperparameters v_{min}, v_{max} and r. The chosen values, for which the network performed the best in the validation set, are $v_{min} = 1$, $v_{max} = 5$ and $r = 0.1$. We average the importance maps for all the subjects of the test set, ending up with a population-based importance map. We use a threshold in order to keep only the top 10% of the voxels with the lowest tolerated noise levels, meaning the most important ones for brain aging. We then apply a gaussian smoothing filter with a kernel value of 1. Different slices of the 3D population-based importance map are shown in Fig. 2. As it can be seen, the areas that are the most relevant for brain aging according to the model's predictions are mesial temporal structures including the hippocampus, brainstem, periventricular and central areas.

4   Discussion

Understanding the logic behind a model's decision is very important in assessing trust and therefore in the adoption of the model by the users [9]. For instance, the users should be assured that correct predictions are an outcome of proper problem representation and not of the mistaken use of artifacts. For this reason, some sort of interpretation may be essential in some cases. In the medical domain [12], the ability of a model to explain the reasoning behind its predictions is an even more important property of a model, as crucial decisions about patient management may need to be made based on its predictions.

In this work, we explored which parts of the brain are important for aging. In order to do so, we made the assumption that unimportant voxels/parts of the brain are not useful for brain age estimation and are not utilized by the prediction model. We trained a prediction model, which accurately estimated brain aging, and a noise model, whose purpose is to increase the noise in the input images voxel-wise, while also keeping the performance of the prediction model unaffected. As can be seen from Fig. 2, our importance maps are in agreement with the existing neuropathology literature [23]. More specifically, it is shown that the hippocampus and parts of the ventricles are where the prediction model focuses to make its decisions.

On the other hand, the differences in the cerebral cortex appear to not be getting captured by the network. In our understanding, there are two reasons behind this. Firstly, the age range of the subjects (44–73 years old) is not large enough for the network to make conclusions. At the same time, the changes in the cerebral cortex are more noticeable after the age of 65 years. In our case, we probably do not have enough subjects in that age range in order to facilitate the network into capturing these differences.

The images that have been used in this study are non-linearly registered, since UKBB provides them ready for use and in the literature more works use the provided preprocessed dataset and therefore, comparison is easier to be done. However, it has been noticed that using non-linearly registered images may lead to the network's missing of subtle changes away from the ventricles, such as cortical changes [7].

In the future, a similar experiment will be conducted with linearly registered images instead of non-linearly registered ones because we believe that, although the performance of the prediction model might be slightly lower, the importance maps will not only be focused on the ventricles and the hippocampus, but also on more subtle changes in the cerebral cortex. Additionally, UKBB provides a variety of other non-imaging features, including biomedical and lifestyle measures, and we intend to test our method on related regression and classification tasks, such as sex classification. In the case of classification tasks we will be also comparing against gradient-based interpretability approaches, such as Grad-CAM [24] and guided backpropagation [27], since the setting allows for their use.

5　Conclusion

In this work, we extend the use of U-noise [14] for 3D inputs and brain age regression. We use 3D brain MR images to train a prediction model for brain age and we investigate the parts of the brain that play the most important role for this prediction. In order to do so, we implement a noise model, which aims to add as much noise as possible in the input image, without affecting the performance of the prediction model. We then localise the most important regions for the task, by finding the voxels that are the least tolerant to the addition of noise, which for the task of brain age estimation are mesial temporal structures including the hippocampus and periventricular areas. Moving forward, we plan to test our interpretability method on classification tasks, such as sex classification as well, and compare with gradient-based methods, which are valid for such tasks.

Acknowledgements. KMB would like to acknowledge funding from the EPSRC Centre for Doctoral Training in Medical Imaging (EP/L015226/1).

References

1. Alam, S.B., Nakano, R., Kamiura, N., Kobashi, S.: Morphological changes of aging brain structure in MRI analysis. In: 2014 Joint 7th International Conference on Soft Computing and Intelligent Systems (SCIS 2014) and 15th International Symposium on Advanced Intelligent Systems (ISIS 2014), pp. 683–687. IEEE, December 2014. https://doi.org/10.1109/SCIS-ISIS.2014.7044901, http://ieeexplore.ieee.org/document/7044901/
2. Alfaro-Almagro, F., Jenkinson, M., Bangerter, N.K., et al.: Image processing and quality control for the first 10,000 brain imaging datasets from UK Biobank. NeuroImage **166**, 400–424 (2 2018). https://doi.org/10.1016/j.neuroimage.2017.10.034, https://www.sciencedirect.com/science/article/pii/S1053811917308613
3. Alzheimer's Association: 2019 Alzheimer's disease facts and figures includes a special report on Alzheimer's detection in the primary care setting: connecting patients and physicians. Tech. rep. (2019), https://www.alz.org/media/Documents/alzheimers-facts-and-figures-2019-r.pdf

4. Ballester, P.L., da Silva, L.T., Marcon, M., et al.: Predicting brain age at slice level: convolutional neural networks and consequences for interpretability. Front. Psychiatr. **12** (2021). https://doi.org/10.3389/fpsyt.2021.598518, https://doi.org/10.3389/fpsyt.2021.598518/full

5. Bass, C., da Silva, M., Sudre, C., et al.: ICAM-reg: Interpretable classification and regression with feature attribution for mapping neurological phenotypes in individual scans. IEEE Transa. Med. Imag. 2021 (2021), arXiv:2103.02561

6. Bintsi, K.M., Baltatzis, V., Kolbeinsson, A., Hammers, A., Rueckert, D.: Patch-based brain age estimation from MR images. In: Lecture Notes in Computer Science (including subseries Lecture Notes in Artificial Intelligence and Lecture Notes in Bioinformatics). LNCS, vol. 12449, pp. 98–107. Springer Science and Business Media Deutschland GmbH, October 2020. https://doi.org/10.1007/978-3-030-66843-3_10, https://doi.org/10.1007/978-3-030-66843-3_10

7. Dinsdale, N.K., Bluemke, E., Smith, S.M., et al.: Learning patterns of the ageing brain in MRI using deep convolutional networks. NeuroImage **224**, 117401 (2021). https://doi.org/10.1016/j.neuroimage.2020.117401

8. Franke, K., Gaser, C.: Ten years of brainage as a neuroimaging biomarker of brain aging: what insights have we gained? (2019). https://doi.org/10.3389/fneur.2019.00789, www.frontiersin.org

9. Gilpin, L.H., Bau, D., Yuan, B.Z., Bajwa, A., Specter, M., Kagal, L.: Explaining explanations: an overview of interpretability of machine learning. In: Proceedings - 2018 IEEE 5th International Conference on Data Science and Advanced Analytics (DSAA 2018), pp. 80–89 (2019). https://doi.org/10.1109/DSAA.2018.00018, https://arxiv.org/pdf/1806.00069.pdf

10. Good, C.D., Johnsrude, I.S., Ashburner, J., Henson, R.N., Friston, K.J., Frackowiak, R.S.: A voxel-based morphometric study of ageing in 465 normal adult human brains. NeuroImage **14**(1 I), 21–36 (2001). https://doi.org/10.1006/nimg.2001.0786

11. He, K., Zhang, X., Ren, S., Sun, J.: Deep residual learning for image recognition. In: Proceedings of the IEEE Computer Society Conference on Computer Vision and Pattern Recognition, vol. 2016-Decem, pp. 770–778. IEEE Computer Society, December 2016. https://doi.org/10.1109/CVPR.2016.90, http://image-net.org/challenges/LSVRC/2015/

12. Holzinger, A., Biemann, C., Pattichis, C.S., Kell, D.B.: What do we need to build explainable AI systems for the medical domain? arXiv preprint (2017). arXiv:1712.09923

13. Kingma, D.P., Ba, J.L.: Adam: a method for stochastic optimization. In: 3rd International Conference on Learning Representations (ICLR 2015) - Conference Track Proceedings (2015)

14. Koker, T., Mireshghallah, F., Titcombe, T., Kaissis, G.: U-noise: learnable noise masks for interpretable image segmentation. arXiv preprint (2021). arXiv:2101.05791

15. Levakov, G., Rosenthal, G., Shelef, I., Raviv, T.R., Avidan, G.: From a deep learning model back to the brain-Identifying regional predictors and their relation to aging. Hum. Brain Mapp. **41**(12), 3235–3252, August 2020. https://doi.org/10.1002/hbm.25011

16. Liu, Z., Cheng, J., Zhu, H., Zhang, J., Liu, T.: Brain age estimation from mri using a two-stage cascade network with ranking loss. In: Martel, A.L., et al. (eds.) MICCAI 2020. LNCS, vol. 12267, pp. 198–207. Springer, Cham (2020). https://doi.org/10.1007/978-3-030-59728-3_20

17. Pardakhti, N., Sajedi, H.: Brain age estimation based on 3D MRI images using 3D convolutional neural network. Multimedia Tools Appl. **79**(33-34), 25051–25065 (2020). https://doi.org/10.1007/s11042-020-09121-z

18. Paszke, A., Gross, S., Massa, F., et al.: PyTorch: an imperative style, high-performance deep learning library. In: Advances in Neural Information Processing Systems, vol. 32 (2019)

19. Peng, H., Gong, W., Beckmann, C.F., Vedaldi, A., Smith, S.M.: Accurate brain age prediction with lightweight deep neural networks. Med. Image Anal. **68**, 101871 (2021). https://doi.org/10.1016/j.media.2020.101871, https://doi.org/10.1016/j.media.2020.101871

20. Popescu, S.G., Glocker, B., Sharp, D.J., Cole, J.H.: A U-NET model of local brain-age. bioRxiv (2021). https://doi.org/10.1101/2021.01.26.428243, https://doi.org/10.1101/2021.01.26.428243

21. Ronneberger, O., Fischer, P., Brox, T.: U-Net: convolutional networks for biomedical image segmentation. In: Navab, N., Hornegger, J., Wells, W.M., Frangi, A.F. (eds.) MICCAI 2015. LNCS, vol. 9351, pp. 234–241. Springer, Cham (2015). https://doi.org/10.1007/978-3-319-24574-4_28

22. Rumelhart, D.E., Hinton, G.E., Williams, R.J.: Learning representations by back-propagating errors. Nature **323**(6088), 533–536 (1986). https://doi.org/10.1038/323533a0, https://www.nature.com/articles/323533a0

23. Savva, G.M., Wharton, S.B., Ince, P.G., Forster, G., Matthews, F.E., Brayne, C.: Age, Neuropathology, and Dementia. New Engl. J. Med. **360**(22), 2302–2309 (2009). https://doi.org/10.1056/NEJMoa0806142, https://doi.org/10.1056/NEJMoa0806142

24. Selvaraju, R.R., Cogswell, M., Das, A., Vedantam, R., Parikh, D., Batra, D.: Grad-CAM: visual explanations from deep networks via gradient-based localization. Int. J. Comput. Vis. **128**(2), 336–359 (2020). https://doi.org/10.1007/s11263-019-01228-7, https://github.com/

25. Singh, A., Sengupta, S., Lakshminarayanan, V.: Explainable deep learning models in medical image analysis (2020). https://doi.org/10.3390/JIMAGING6060052, www.mdpi.com/journal/jimaging

26. Smilkov, D., Thorat, N., Kim, B., Viégas, F., Wattenberg, M.: SmoothGrad: removing noise by adding noise. arXiv preprint (2017). arXiv:1706.03825

27. Springenberg, J.T., Dosovitskiy, A., Brox, T., Riedmiller, M.: Striving for simplicity: The all convolutional net. In: 3rd International Conference on Learning Representations (ICLR 2015) - Workshop Track Proceedings (2015)

28. Zeiler, M.D., Fergus, R.: Visualizing and understanding convolutional networks. In: Fleet, D., Pajdla, T., Schiele, B., Tuytelaars, T. (eds.) ECCV 2014. LNCS, vol. 8689, pp. 818–833. Springer, Cham (2014). https://doi.org/10.1007/978-3-319-10590-1_53

TDA4MedicalData Workshop

Lattice Paths for Persistent Diagrams

Moo K. Chung[1]([✉]) and Hernando Ombao[2]

[1] University of Wisconsin, Madison, USA
mkchung@wisc.edu
[2] King Abdullah University of Science and Technology, Thuwal, Saudi Arabia
hernando.ombao@kaust.edu.sa

Abstract. Persistent homology has undergone significant development in recent years. However, one outstanding challenge is to build a coherent statistical inference procedure on persistent diagrams. In this paper, we first present a new lattice path representation for persistent diagrams. We then develop a new exact statistical inference procedure for lattice paths via combinatorial enumerations. The lattice path method is applied to the topological characterization of the protein structures of the COVID-19 virus. We demonstrate that there are topological changes during the conformational change of spike proteins.

1 Introduction

Despite its rigorous mathematical foundation developed for two decades starting with study [13], persistent homology still suffers from numerous statistical and computational problems. It has not yet become a standard tool in medical imaging. Persistent homology has been applied to a wide variety of data including brain networks [8], protein structures [15], RNA viruses [5] and molecular structures [19]. However, most of these methods only serve as exploratory tools that provide descriptive summary statistics rather than formal inference. The main difficulty is due to the heterogeneous nature of topological features, which do not have a one-to-one correspondence across persistent diagrams. Motivated by these challenges, we propose a more principled topological inference procedure through lattice paths.

Lattice paths are widely studied algebraic objects in combinatorics and may have potential applications in persistent homology [2,8,21,23]. Here, we propose to use the lattice path approach in computing probabilistic statements about the similarity of two persistent diagrams. This is often needed to produce some baseline quantitive measure, such as the p-value, commonly used in biomedical research [7,8]. Existing methods for computing p-values usually rely on approximate time consuming resampling techniques: jackknife, bootstrap and the permutation test [1,8]. However, our approach is analytic and thus computes the <u>exact</u> probability without computational burden.

The main contributions of this paper are the following: (1) a new data representation via Dyck and lattice paths; (2) the analytic approach for computing

M. Reyes et al. (Eds.): iMIMIC 2021/TDA4MedicalData 2021, LNCS 12929, pp. 77–86, 2021.
https://doi.org/10.1007/978-3-030-87444-5_8

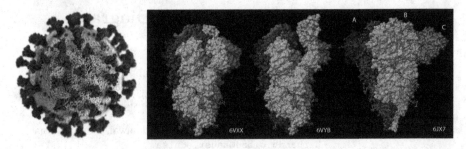

Fig. 1. Left: COVID-19 virus with spike proteins (red). Right: Spike proteins of the three different corona viruses. The spike proteins consist of three similarly shaped interwinding substructures identified as A (blue), B (red) and C (green) domains. (Color figure online)

probabilities without resampling and significantly reducing run time; (3) the first topological study on the shape of COVID-19 virus spikes proteins. The proposed lattice path method was used in differentiating the conformational changes of the COVID-19 virus spike proteins that is needed for the virus to penetrate host cells (Fig. 1). This demonstration is particularly relevant due to the potential for advancing vaccine development and the current public health concern [4,25].

2 Methods

Simplicial Homology. High dimensional objects, such as proteins and molecules, can be modeled as a point cloud data V consisting of p number of points (atoms) indexed as $V = \{1, 2, \cdots, p\}$. Suppose that the distance ρ_{ij} between points i and j satisfies the metric properties. For proteins, we can simply use the Euclidean distance between atoms in a molecule. Then $\mathcal{X} = (V, \rho), \rho = (\rho_{ij})$ is a metric space where we can build a filtration necessary for persistent homology. If we connect points following some criterion on the distance, they will form a simplicial complex which will follow the topological structure of the molecule [12,18,28]. The k-simplex is the convex hull of $k + 1$ points in V. A simplicial complex is a finite collection of simplices such as points (0-simplex), lines (1-simplex), triangles (2-simplex) and higher dimensional counterparts. In particular, the *Rips complex* $\mathcal{X}_\epsilon$ is a simplicial complex, whose k-simplices are formed by $(k+1)$ points which are pairwise within distance ϵ [16]. The Rips complex induces a hierarchical nesting structure called the Rips filtration $\mathcal{X}_{\epsilon_0} \subset \mathcal{X}_{\epsilon_1} \subset \mathcal{X}_{\epsilon_2} \subset \cdots$ for filtration values $0 = \epsilon_0 < \epsilon_1 < \epsilon_2 < \cdots$. The filtration is quantified through k-*cycles* where 0-cycles are the connected components, 1-cycles are loops while 2-cycles are 3-simplices (tetrahedron) without interior. During the Rips filtration, the i-th k-cycles are born at filtration value b_i and die at d_i. The collection of all the paired filtration values $\{(b_1, d_1), \cdots, (b_q, d_q)\}$ displayed as scatter points in 2D plane is called the *persistent diagram*.

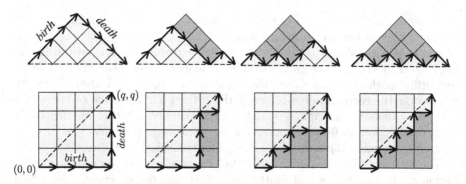

Fig. 2. Top: 4 different Dyck paths out of 14 possible paths for $q = 4$. Bottom: corresponding lattice paths.

Dyck Paths. The first step in the proposed *lattice path method* is to sort the set of all the birth and death values in the filtration as order statistics $c : c_{(1)} < c_{(2)} < \cdots < c_{(2q)}$, where $c_{(i)}$ is one of the birth or death values. The subscript (i) denotes the i-th smallest value. We will simply call such sequence as the *birth-death process*. Every possible valid sequence of birth and death values can be viewed as forming a probability space, where each valid sequence is likely to happen with equal probability. During the filtration, the sequence of birth and death occurs somewhat randomly but still maintains a specific pairing structure.

There exists a one-to-one relation between the ordering information and Dyck paths if we identify births with $\nearrow$ and deaths with $\searrow$ [2,21]. If we trace the arrows, we obtain the Dyck path (Fig. 2) [23]. A valid Dyck path always starts at $y = 0$ and ends at $y = 0$. At any moment during the filtration, a Dyck path cannot go below $y = 0$. The total number of Dyck paths is called the Catalan number $\kappa_p = \frac{1}{q+1}\binom{2q}{q}$. The first few Catalan numbers are $\kappa_1 = 1, \kappa_2 = 2, \kappa_3 = 5$ and $\kappa_4 = 14$. More rapid changes in the direction of Dyck paths imply more fleeting fluctuations which are indicative of smaller topological signals. Less fluctuations indicate larger persistence and thus larger topological structures. The first path in Fig. 2 has larger persistence while the last path has smaller persistence.

Lattice Paths. If we rotate the Dyck paths clockwise at 45° and flip vertically, we obtain equivalent *monotone lattice paths* consisting of a sequence of $\rightarrow$ (uparrow) and $\uparrow$ (downarrow). Figure 2 displays corresponding monotone *lattice paths* between $(0,0)$ and (q,q) where the path does not pass above the diagonal line $y = x$ [23]. During the filtration, there cannot be more deaths than births and thus the path must lie below the diagonal line. The total area below Dyck paths can be used to quantify the Dyck paths [6]. Since the area below a Dyck path is equivalent to $q^2/2$ subtracted by the total area of boxes below the corresponding lattice path, we will simply use lattice paths for quantification. If we tabulate how the area of boxes change over the x-coordinate in a lattice path, it is monotone. In the first path in Fig. 2, the number of boxes below the first and the last lattice paths are $(0,0,0,0)$ and $(0,1,2,3)$. The area below the path is related to persistence. A barcode with smaller persistences (last path in Fig. 2) will have more boxes (dark gray boxes) while longer persistences will have fewer boxes (first path in Fig. 2). Given the sequence of heights of piled-up boxes, we can recover the corresponding lattice path by tracing the outline of boxes. We can further recover the original pairing information about births and deaths. In the Rips filtration for 0-cycles, persistent diagrams line up vertically as $(0, d_{(i)})$. We simply augment them as $((i-1)\delta, d_{(i)})$ for sufficiently small δ.

The lattice and Dyck path representations only encode the ordering information about how births and deaths are paired, and do not encode the actual filtration values. This is remedied by adaptively weighting the length of arrows in lattice paths. We sort the set of birth values b_i and death values d_i as the order statistics:

$$b_{(1)} < b_{(2)} < \cdots < b_{(q-1)} < b_{(q)}, \quad d_{(1)} < d_{(2)} < \cdots < d_{(q-1)} < d_{(q)}.$$

We start at origin $(0,0)$. When we encounter a birth $b_{(i)}$, we take the horizontal step to $b_{(i)}$. When we encounter a death $d_{(i)}$, take the vertical step to $d_{(i)}$ (Fig. 2). The weighted lattice paths contains the same topological information as the original persistent diagram.

Exact Topological Inference. Using the weighted lattice paths, we can provide the probabilistic statement about the discrepancy between two birth-death processes which can be used for topological inference. For this, we need the transformation ϕ:

Theorem 1. *There exists a one-to-one map from a birth-death process to a monotone function ϕ with $\phi(0) = 0$ and $\phi(1) = q$.*

Proof. We explicitly construct such a function ϕ. Consider the sequence of areas of boxes as we traverse the weighed lattice path: $h : h_1 \leq h_2 \leq \cdots \leq h_q$, where $h_{i+1} = (b_{(i+1)} - b_{(i)})(d_{(i+1)} - d_{(i)})$ is the area of i-th box with $h_1 = 0$. The areas h may not strictly increase (Fig. 3). If births occurs r times sequentially in the birth-death process, h will have r repeated identical areas $h_i, \cdots, h_i$ as a subsequence. To make the subsequence strictly increasing, we simply add a sequence of strictly increasing small numbers $\delta(0, 1, 2, \cdots, r-1)$ to the repetition with $\delta \leq \frac{1}{r}$ (Fig. 3). Denote the transformed sequence as $h' : h'_1 < \cdots < h'_q$. Then $\phi(t)$ is given as a step function

$$\phi(t) = \begin{cases} 0 & \text{if } t \in [0, \frac{h'_1}{q}) \\ j & \text{if } t \in [\frac{h'_j}{q}, \frac{h'_{j+1}}{q}) \text{ for } j = 1, \cdots, q-1 \\ q & \text{if } t \in [\frac{h'_q}{q}, 1] \end{cases}.$$

From $\phi(t)$, the original sequence h and the original birth-death process can be recovered exactly. Such a map from a birth-death process to ϕ is one-to-one. $\square$

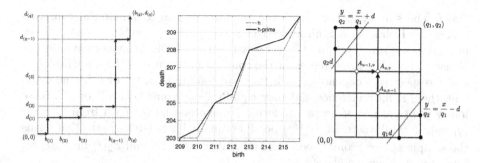

Fig. 3. Left: Weighted lattice path equivalent to a persistent diagram. Middle: The area of boxes below lattice paths h (dotted line) is made into strictly increasing to h' (solid line). Right: The problem of lattice path enumeration between $(0,0)$ and (q_1, q_2) with the constraint $|x/q_1 - y/q_2| < d$.

Note $\|h - h'\|_2 \to 0$ as $\delta \to 0$. So by making δ as small as possible, we can construct a strictly monotone h' to be arbitrarily close to h. The normalized step function $\phi(t)/q$ can be viewed as an *empirical cumulative distribution* and many statistical tools for analyzing distributions can be readily applied. Figure 4-bottom displays the lattices paths and the normalized step functions of 1-cycles corresponding to the spike proteins used in the study.

With monotone function ϕ, we are ready to test the *topological equivalence* of two birth-death processes:

$$C^1 : c_1^1 < c_2^1 < \cdots < c_{q_1}^1, \quad C^2 : c_1^2 < c_2^2 < \cdots < c_{q_2}^2.$$

Let ϕ_1 and ϕ_2 be the step functions corresponding to C^1 and C^2. The topological distance

$$D(\phi_1, \phi_2) = \sup_{t \in [0,1]} \left| \frac{\phi_1(t)}{q_1} - \frac{\phi_2(t)}{q_2} \right|$$

will be used as the test statistic for testing the equivalence of C^1 and C^2. The normalizing denominators q_1 and q_2 ensures that the value of step functions are in $[0,1]$. The statistic $D(\phi_1, \phi_2)$ is the upper bound of area difference under $\phi_1(t)/q_1$ and $\phi_2(t)/q_2$:

$$\int_0^1 \left| \frac{\phi_1(t)}{q_1} - \frac{\phi_2(t)}{q_2} \right| dt \leq D(\phi_1, \phi_2).$$

Theorem 2. *Under the null hypothesis of equivalence of C^1 and C^2,*

$$P(D(\phi^1, \phi^2) \geq d) = 1 - \frac{A_{q_1, q_2}}{\binom{q_1 + q_2}{q_1}},$$

where $A_{u,v}$ satisfies $A_{u,v} = A_{u-1,v} + A_{u,v-1}$ with the boundary condition $A_{q_1, 0} = A_{0, q_2} = 1$ within the band $|u/q_1 - v/q_2| < d$.

Proof. The statement can be proved similarly as the combinatorial construction of the Kolmogorov-Smirnov test [3,8,17]. First, we combine monotonically increasing sequences C^1 and C^2 and sort them into a bigger monotone sequence of size $q_1 + q_2$. Then, we represent the combined sequence as the sequence of $\rightarrow$ and $\uparrow$ respectively depending on if they are coming from C^1 or C^2. Under the null, there is no preference and they equally likely come from C^1 or C^2. If we follow the sequence of arrows, it forms a monotone lattice path from $(0,0)$ to (q_1, q_2). In total, there are $\binom{q_1+q_2}{q_1}$ possible equally likely lattice paths that forms the sample space. From Theorem 1, the values of $\phi_1(t)$ and $\phi_2(t)$ are integers from 0 to q. Then it follows that

$$P(D \geq d) = 1 - P(D_q < d) = 1 - \frac{A_{q_1,q_2}}{\binom{q_1+q_2}{q_1}},$$

where $A_{u,v}$ is the total number of valid paths from $(0,0)$ to (u,v) within dotted red lines in Fig. 3. $A_{u,v}$ is iteratively computed using $A_{u,v} = A_{u-1,v} + A_{u,v-1}$. with the boundary condition $A_{u,0} = A_{0,v} = 1$ for all u and v. $\qquad\square$

Computing A_{q_1,q_2} iteratively requires at most $q_1 \cdot q_2$ operations while the permutation test will cause a computational bottleneck for large q_1 and q_2. Thus, the proposed *lattice path method* computes the exact p-value substantially faster than the permutation test. Since most protein molecules consist of thousands of atoms, q_1 and q_2 should be sufficiently large to apply the asymptotic [10,17,22]:

Theorem 3. $\lim_{q_1,q_2 \to \infty} P\left(\sqrt{\frac{q_1 q_2}{q_1+q_2}} D \geq d\right) = 2\sum_{j=1}^{\infty}(-1)^{j-1}e^{-2j^2 d^2}.$

Subsequently, the p-value under the null hypothesis is given by

$$p\text{-value} = 2e^{-d_o^2} - 2e^{-8d_o^2} + 2e^{-18d_o^2}\cdots,$$

where d_o is the observed value of $\sqrt{\frac{q_1 q_2}{q_1+q_2}}D$. Computing the p-value through Theorem 3 mainly requires sorting, which has the runtime of $\mathcal{O}(q \log q)$ for $q = q_1 = q_2$. On the other hand, the traditional permutation test requires computing the distance for $\binom{2q}{q}$ possible permutations, which is asymptotically $\mathcal{O}(4^q/\sqrt{\pi q})$ [11,14]. For thousands of atoms, the total number of permutations is too large to compute. Thus, only a small fraction of randomly generated permutations are used in the traditional permutation test [9–11,20,24,27]. Even if we use hundreds of thousands permutations, the traditional permutation test still takes a significant computational effort. Further, as an approximation procedure, the standard permutation test does not perform better than the exact topological inference, which gives the mathematical ground truth. This is demonstrated in Table 1 in the simulation study [9].

3   Application: Spike Proteins of COVID-19 Virus

The proposed lattice path method is used to study the topological structure of the severe acute respiratory syndrome coronavirus 2 (SARS-Cov-2), which is

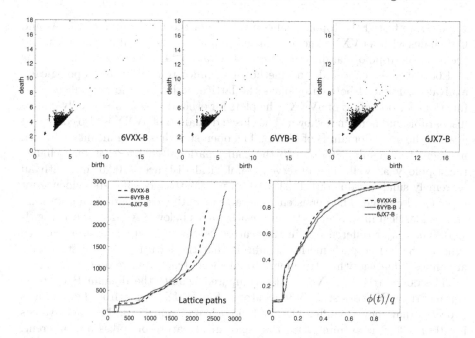

Fig. 4. Top: persistent diagrams of three different spike proteins. The red dots are 0-cycles and the black dots are 1-cycles. The units are in Å (angstrom). Bottom: the corresponding lattice paths and normalized step functions $\phi(t)/q$. (Color figure online)

often called COVID-19. Since the start of the global pandemic (approximately December 2019), COVID-19 has already caused 3.85 million deaths in the world as of June 2021. The COVID-19 virus is specific member of a much broader coronavirus family, which all have spike proteins surrounding the spherically shaped virus similar to the sun's corona. The glycoprotein spikes bind with receptors on the host's cells and consequently cause severe infection. The atomic structure of spike proteins can be determined through the cryogenic electron microscopy (cryo-EM) [4,25]. Figure 1-left illustrates spike proteins (colored red) that surround the spherically shaped virus. Each spike consists of three similarly shaped protein molecules with rotational symmetry often identified as A, B and C domains. The spike proteins have two distinct conformations identified as *open* and *closed* states, where the domain's opening is necessary for interfacing with the host cell's surface for membrane fusion and viral entry (Fig. 1-right). Indeed, most current vaccine efforts focus on preventing the open state from interfacing with the host cell. Hence, this line of research is of prime importance in vaccine development and therapeutics [4].

In this study, we analyzed the spikes of three different coronaviruses identified as 6VXX, 6VYB [25] and 6JX7 [26]. The 6VXX and 6VYB are respectively the closed and open states of SARS-Cov-2 from human while 6JX7 is feline coronavirus (Fig. 1). All the domains of 6VXX have exactly 7604 atoms and are expected to be topologically identical. Applying the lattice path method to

1-cycles, we tested the topological equivalence of the B-domain and the A- and C-domains within 6VXX. The normalized step functions are almost identical and the observed topological distances are 0.0090 and 0.0936, which give the p-value of 1.00 each. As expected, the method concludes that they are topologically equivalent. Figure 4-bottom displays the lattice paths and the normalized step functions for domain B of 6VXX. The plots for other domains are visually almost inseparable and hence not shown. The closed domain B of 6VXX is also compared against the open domain B of 6VYB. The open state has a significantly reduced number of atoms at 6865 due to the conformational change that may change the topology as well. The observed topological distance is 0.20 and with an extremely small p-value of 8.1123×10^{-38} which strongly suggests evidence for topological change. The persistent diagrams of both closed and open states are almost identical in smaller birth and death values below 6 Å (angstrom) (Fig. 4-top). The major difference is in the scatter points with larger birth and death values. The lattice path method confirms that the local topological structures are almost identical while the global topological structures are different.

The domain B of 6VXX is also compared against the domain B of feline coronavirus 6JX7 consisting of 9768 atoms. Since 6JX7 is not from human, it is expected that they are different. The topological distance is 0.9194 and p-values is 0.00×10^{-38} confirming that the topological nature of spikes are different. This shows the biggest difference among all the comparisons done in this study. The Matlab codes and data used for the study are available at http://www.stat.wisc.edu/~mchung/TDA.

4   Conclusions

In this paper, we proposed a new representation of persistent diagrams using lattice paths. The novel representation enable us to perform the statistical inference combinatorially by enumerating every possible valid lattice paths analytically. The proposed lattice path method is subsequently used to analyze the coronavirus spike proteins. The normalized step functions $\phi(t)/q$ for all the spike proteins show fairly stable consistent global monotone pattern but with localized differences. We demonstrated the lattice path method has the ability to statistically discriminate between the conformational changes of the spike protein that are needed in the transmission of the virus. We hope that the our new representation enables scientists in their effort to automatically identify the different types and states of coronaviruses in a more principled manner.

Acknowledgement. The illustration of COVID-19 virus (Fig. 1 left) is provided by Alissa Eckert and Dan Higgins of Disease Control and Prevention (CDC), US. The proteins 6VXX and 6VYB are provided by Alexander Walls of University of Washington. The protein 6JX7 is provided by Tzu-Jing Yang of National Taiwan University. Figure 2-left is modified from an image in Wikipedia. This study is supported by NIH EB022856 and EB028753, NSF MDS-2010778, and CRG from KAUST.

References

1. Ahmed, M., Fasy, B.T., Wenk, C.: Local persistent homology based distance between maps. In: Proceedings of the 22nd ACM SIGSPATIAL International Conference on Advances in Geographic Information Systems, pp. 43–52 (2014)
2. Billera, L.J., Holmes, S.P., Vogtmann, K.: Geometry of the space of phylogenetic trees. Adv. Appl. Math. **27**, 733–767 (2001)
3. Böhm, W., Hornik, K.: A Kolmogorov-Smirnov test for r samples. Inst. Stat. Math. Res. Rep. Ser. Rep. **117**, 105 (2010)
4. Cai, Y., et al.: Distinct conformational states of SARS-CoV-2 spike protein. Science **369**, 1586–1592 (2020)
5. Chan, J.M., Carlsson, G., Rabadan, R.: Topology of viral evolution. Proc. Nat. Acad. Sci. **110**(46), 18566–18571 (2013)
6. Chapman, R.: Moments of Dyck paths. Discrete Math. **204**, 113–117 (1999)
7. Chazal, F., Fasy, B.T., Lecci, F., Rinaldo, A., Singh, A., Wasserman, A.: On the bootstrap for persistence diagrams and landscapes. arXiv preprint arXiv:1311.0376 (2013)
8. Chung, M.K., Lee, H., DiChristofano, A., Ombao, H., Solo, V.: Exact topological inference of the resting-state brain networks in twins. Network Neurosci. **3**, 674–694 (2019)
9. Chung, M.K., Luo, Z., Leow, A.D., Alexander, A.L., Davidson, R.J., Hill Goldsmith, H.: Exact combinatorial inference for brain images. In: Frangi, A.F., Schnabel, J.A., Davatzikos, C., Alberola-López, C., Fichtinger, G. (eds.) MICCAI 2018. LNCS, vol. 11070, pp. 629–637. Springer, Cham (2018). https://doi.org/10.1007/978-3-030-00928-1_71
10. Chung, M.K., Villalta-Gil, V., Lee, H., Rathouz, P.J., Lahey, B.B., Zald, D.H.: Exact topological inference for paired brain networks *via* persistent homology. In: Niethammer, M., et al. (eds.) IPMI 2017. LNCS, vol. 10265, pp. 299–310. Springer, Cham (2017). https://doi.org/10.1007/978-3-319-59050-9_24
11. Chung, M.K., Xie, L., Huang, S.-G., Wang, Y., Yan, J., Shen, L.: Rapid acceleration of the permutation test via transpositions. In: Schirmer, M.D., Venkataraman, A., Rekik, I., Kim, M., Chung, A.W. (eds.) CNI 2019. LNCS, vol. 11848, pp. 42–53. Springer, Cham (2019). https://doi.org/10.1007/978-3-030-32391-2_5
12. Edelsbrunner, H., Harer, J.: Computational Topology: An Introduction. American Mathematical Society (2010)
13. Edelsbrunner, H., Letscher, D., Zomorodian, A.: Topological persistence and simplification. In: Proceedings. 41st Annual Symposium on Foundations of Computer Science, pp. 454–463. IEEE (2000)
14. Feller, W.: An Introduction to Probability Theory and its Applications, vol. 2. John Wiley & Sons, Hoboken (2008)
15. Gameiro, M., Hiraoka, Y., Izumi, S., Kramar, M., Mischaikow, K., Nanda, V.: A topological measurement of protein compressibility. Jpn. J. Ind. Appl. Math. **32**(1), 1–17 (2014). https://doi.org/10.1007/s13160-014-0153-5
16. Ghrist, R.: Barcodes: the persistent topology of data. Bull. Am. Math. Soc. **45**, 61–75 (2008)
17. Gibbons, J.D., Chakraborti, S.: Nonparametric Statistical Inference. Chapman & Hall/CRC Press, Boca Raton (2011)
18. Hart, J.C.: Computational topology for shape modeling. In: Proceedings of the International Conference on Shape Modeling and Applications, pp. 36–43 (1999)

19. Meng, Z., Xia, K.: Persistent spectral-based machine learning (PerSpect ML) for protein-ligand binding affinity prediction. Sci. Adv. **7**(19), eabc5329 (2021)
20. Nichols, T.E., Holmes, A.P.: Nonparametric permutation tests for functional neuroimaging: a primer with examples. Hum. Brain Map. **15**, 1–25 (2002)
21. Simion, R.: Noncrossing partitions. Discrete Math. **217**(1–3), 367–409 (2000)
22. Smirnov, N.V.: Estimate of deviation between empirical distribution functions in two independent samples. Bull. Moscow Univ. **2**, 3–16 (1939)
23. Stanley, R.P.: Enumerative combinatorics. In: Cambridge Studies in Advanced Mathematics, vol. 2 (1999)
24. Thompson, P.M., et al.: Genetic influences on brain structure. Nat. Neurosci. **4**, 1253–1258 (2001)
25. Walls, A.C., Park, Y.-J., Tortorici, M.A., Wall, A., McGuire, A.T., Veesler, D.: Structure, function, and antigenicity of the SARS-CoV-2 spike glycoprotein. Cell **181**, 281–292 (2020)
26. Yang, Z., Wen, J., Davatzikos, C.: Smile-GANs: semi-supervised clustering via GANs for dissecting brain disease heterogeneity from medical images. arXiv preprint arXiv:2006.15255 (2020)
27. Zalesky, A., et al.: Whole-brain anatomical networks: does the choice of nodes matter? NeuroImage **50**, 970–983 (2010)
28. Zomorodian, A.J.: Topology for Computing. Cambridge Monographs on Applied and Computational Mathematics, vol. 16. Cambridge University Press, Cambridge (2009)

Neighborhood Complex Based Machine Learning (NCML) Models for Drug Design

Xiang Liu[1,2,3] and Kelin Xia[1(✉)]

[1] Division of Mathematical Sciences, School of Physical and Mathematical Sciences, Nanyang Technological University, Singapore 637371, Singapore
`xiakelin@ntu.edu.sg`
[2] Chern Institute of Mathematics and LPMC, Nankai University, Tianjin 300071, China
[3] Center for Topology and Geometry Based Technology, Hebei Normal University, Hebei 050024, China

Abstract. The importance of drug design cannot be overemphasized. Recently, artificial intelligence (AI) based drug design has begun to gain momentum due to the great advancement in experimental data, computational power and learning models. However, a major issue remains for all AI-based learning models is efficient molecular representations. Here we propose Neighborhood complex (NC) based molecular featurization (or feature engineering), for the first time. In particular, we reveal deep connections between NC and Dowker complex (DC) for molecular interaction based bipartite graphs, for the first time. Further, NC-based persistent spectral models are developed and the associated persistent attributes are used as molecular descriptors or fingerprints. To test our models, we consider protein-ligand binding affinity prediction. Our NC-based machine learning (NCML) models, in particular, NC-based gradient boosting tree (NC-GBT), are tested on three most-commonly used datasets, i.e., including PDBbind-v2007, PDBbind-v2013 and PDBbind-v2016, and extensively compared with other existing state-of-the-art models. It has been found that our NCML models can achieve state-of-the-art results.

Keywords: Neighborhood complex · Machine learning · Dowker complex · Hodge Laplacian · Drug design

1 Introduction

Featurization or feature engineering is of essential importance for artificial intelligence (AI) based drug design. The performance of quantitative structure activity/property relationship (QSAR/QSPR) models and machine learning models for biomolecular data analysis is largely determined by the design of proper molecular descriptors/fingerprints. Currently, more than 5000 types of molecular

© Springer Nature Switzerland AG 2021
M. Reyes et al. (Eds.): iMIMIC 2021/TDA4MedicalData 2021, LNCS 12929, pp. 87–97, 2021.
https://doi.org/10.1007/978-3-030-87444-5_9

descriptors, which are based on molecular structural, chemical, physical and biological properties, have been proposed [17,23]. Among these molecular features,
structural descriptors are the most-widely used ones. Further, various molecular
fingerprints are proposed, Different from molecular descriptors, molecular fingerprint is large-sized vector of molecular features that are systematically generated
based on molecular properties, in particular, structural properties. Deep learning
models, such as antoencoder, CNN, and GNN, have also been used in molecular
fingerprint generation [7,31].

The transferability of machine learning models is highly related to molecular descriptors or fingerprints. Features that characterize more intrinsic and
fundamental properties can be better shared between data and "understood"
by machine learning models. Mathematical invariants, from geometry, topology,
algebra, combinatorics and number theory, are highly abstract quantities that
describe the most intrinsic and fundamental rules and properties in nature sciences. Mathematical invariants, in particular, topological and geometric invariants, based molecular descriptors have achieved great successes in various steps
of drug design [3,4,20], including protein-ligand binding affinity prediction, protein stability change upon mutation prediction, toxicity prediction, solvation free
energy prediction, partition coefficient and aqueous solubility, binding pocket
detection, and drug discovery. These models have also demonstrated great advantages over traditional molecular representations in D3R Grand challenge [21].

Neighborhood complex (NCs) was originally introduced by Lovász in his
proof of Kneser's conjecture [18], and was used to provide some of the first
nontrivial algebra-topological lower bounds for chromatic numbers of graphs
[13,14], which laid the foundations of topological combinatorial by introducing
homotopy theoretical methods to combinatorics. For any graph, if its NC is
k-connected, then the chromatic number of the graph is at least $k + 3$. Mathematically, a graph based NC is a simplicial complex formed by the subsets of the
neighborhood of vertices. Essentially, each vertex has a set of neighbors (vertices
connected through edges), and neighborhood simplexes are generated among the
subsets of these neighborhood vertices. These vertex-neighbor-based simplexes
combine together to form the neighborhood complex for the graph. Dowker complex (DC) was developed for the characterization of relations between two sets
[6]. Essentially, two different Dowker complexes can be constructed from the
relations between the two different sets. These two DCs share not only the same
homology groups, but also the homotopy groups [5].

Here we propose neighborhood complex based machine learning (NCML) for
drug design, for the first time. Neighborhood complex is applied for molecular structure and interaction characterization. In particular, we have revealed
that when molecular interactions are represented as a bipartite graph, the corresponding two individual NCs are exactly two DCs that share the same homology groups. More specifically, protein-based NC/DC and ligand-based NC/DC,
which are generated from the protein-ligand bipartite graph, have the same Betti
number. Further, we construct NC-based persistent spectral models and use their
persistent attributes as molecular descriptors. Our NC-based machine learning

models, in particular, NC-based gradient boosting tree (NC-GBT), are extensively tested on the three most-commonly used datasets from the well-established protein-ligand binding databank, i.e., PDBbind. It is found that our NC-GBT model can achieve competitive results.

2 Method

2.1 NC-Based Biomolecular Structure and Interaction Analysis

Molecular representation and featurization are of essential importance for the analysis of molecular data from materials, chemistry and biology. Mathematical invariant based molecular descriptors are of greater transferability thus have achieved better performance in AI-based drug design [3,20]. Here we propose the first NC-based representations for molecular structure and interaction analysis.

NC-Based Biomolecular Structure Characterization. Graph theory is widely used for the description and characterization of molecular structures and interactions. A molecular graph $G = (V, E)$ is composed of a set of vertices V, with each vertex representing molecular atom, residue, motif, or even the entire molecule, and a set of edges E, representing interactions of various kinds including covalent bonds, van der Walls, electrostatic, and other non-covalent forces. In comparison with a graph, a simplicial complex has not only vertices and edges, but also has triangles, tetrahedrons and other higher-dimensional counterparts, which characterize more complicated "many-body" interactions (other than just pair-interactions as in graphs). A natural way to generate a simplicial complex from a graph is to consider vertex neighborhood topology. More specifically, for a graph $G = (V, E)$ with $V = \{v_0, v_1, ..., v_n\}$, its neighborhood complex $K_N(G)$ shares the same vertex set V, its k-simplex is formed among any $k + 1$ vertices $\{v_0, v_1, ..., v_k\}$ if they share a common neighborhood vertex, i.e., there exists a vertex $v_j \in V$ such that $\{(v_i, v_j)|i = 0, 1..., k\} \subset E$.

NC-Based Biomolecular Interaction Characterization. Both intra- and inter- molecular interactions, i.e., interactions within and between molecules, can be represented as bipartite graphs (also known as bigraphs or 2-mode networks). Mathematically, a bipartite graph $G(V_1, V_2, E)$ has two vertex sets V_1 and V_2, and all its edges are formed only between the two vertex sets. An NC $K_N(G)$ can be constructed from a bipartite graph $G(V_1, V_2, E)$. Since the edges for the bipartite graph exist only between two vertex sets, the NC $K_N(G)$ is divided into two separated and isolated simplicial complexes $K_{N,1}(G)$ with all the vertices in V_1 and $K_{N,2}(G)$ with all the vertices in V_2. Further, two Dowker complexes $K_{D,1}(G)$ and $K_{D,2}(G)$ can be constructed from a bipartite graph $G(V_1, V_2, E)$. The DC $K_{D,1}(G)$ is defined on V_1, and its k-simplex is formed among $k + 1$ vertices which have "relations", i.e., forming edges, with a common vertex in V_2. Similarly, the DC $K_{D,2}(G)$ is based on V_2, and its k-simplex is formed among $k+1$ vertices which are "related to" a common vertex in V_1. Note that when vertices

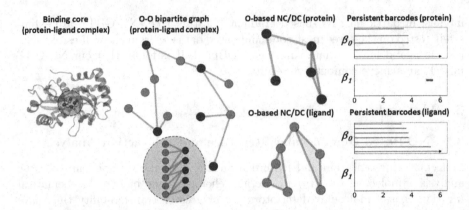

Fig. 1. NC-based representations for O-O pair of PDBID 3FV1. A binding core region with cutoff value 3.5Å is firstly extracted, that is, all the O atoms of ligand and O atoms of protein within 3.5Å from ligand. Then a bipartite graph based filtration can be constructed, and the corresponding NC based filtration can be generated. The NC consist of two disjoint components, one is in protein and the other is in ligand. So the filtration can be divided into two parts. Here we only draw one bipartite graph and its NC components in protein and ligand. Finally, the corresponding persistent barcodes are computed. It can be seen that the persistent barcodes in positive dimensions are totally same for the two filtrations in protein and ligand.

are related to a common vertex, that means they share the same common neighborhood vertex and a NC-based simplex will be formed among them. In fact, for bipartite graphs, the two types of simplicial complexes, i.e., NCs, and DCs, are equivalent to each other, that is $K_{N,1}(G) = K_{D,1}(G)$ and $K_{N,2}(G) = K_{D,2}(G)$. Further, the Dowker theorem states that the homology of $K_{D,1}(G)$ and $K_{D,2}(G)$ are isomorphic, which means $H_p(K_{D,1}(G)) \cong H_p(K_{D,1}(G))(p > 0)$, and 0-th homology is isomorphic if the bipartite graph G is connected [6]. With the connection between NCs and DCs, we have $H_p(K_{N,1}(G)) \cong H_p(K_{N,2}(G))(p > 0)$.

Figure 1 illustrates the bipartite graph-based NC/DCs and their persistent homology. The bipartite graph is generated from the oxygen-oxygen (O-O) atom pairs from the protein-ligand complex 3FV1. The corresponding NC has two disjoint components, one in protein and the other in ligand. The distance-based filtration process is considered (detailed in the section below) and two persistent barcodes, for protein and ligand NCs respectively, are generated. It can be observed that β_1 bars are exactly the same. Note that β_0 bars are not the same as the bipartite graphs are not always connected during the filtration process.

2.2 NC-Based Persistent Spectral Models

For all the persistent models, including persistent homology/cohomology, persistent spectral and persistent function, the key point is the filtration process. For topology-based protein-ligand interaction models, i.e. our bipartite graph based NC models, we can define the filtration parameter f as the Euclidean distance or electrostatic interactions for each edge of the bipartite graph. The electrostatic interaction between two atoms v_i and v_j is defined as $\frac{1}{1+exp(-\frac{cq_iq_j}{d(v_i,v_j)})}$ with q_i and q_j the partial charges of atoms v_i and v_j and parameter c a constant value (usually taken as 100). With the increase of filtration value, a series of nested bipartite graphs can be generated,

$$G_{f_0} \subset G_{f_1} \subset ... \subset G_{f_n}. \tag{1}$$

Here $(f_0 \leqslant f_1 \leqslant ... \leqslant f_n)$ are filtration values and $G_{f_i} (0 \leqslant i \leqslant n)$ is a bipartite graph consists of all the edges whose filtration values are not bigger than f_i. The corresponding NCs can be constructed accordingly as follows,

$$K_N(G_{f_0}) \subset K_N(G_{f_1}) \subset ... \subset K_N(G_{f_n}). \tag{2}$$

In fact, we have two disjointed series of nested NCs as follows,

$$K_{N,1}(G_{f_0}) \subset K_{N,1}(G_{f_1}) \subset ... \subset K_{N,1}(G_{f_n}). \tag{3}$$
$$K_{N,2}(G_{f_0}) \subset K_{N,2}(G_{f_1}) \subset ... \subset K_{N,2}(G_{f_n}). \tag{4}$$

Note that the first NC series $\{K_{N,1}(G_{f_i})\}$ are for protein part, all of their vertices are protein atoms. In contrast, the second NC series $\{K_{N,2}(G_{f_i})\}$ are fully based on ligand atoms. From Dowker's theorem, these two NC series share the same homology groups, i.e., $H_p(K_{N,1}(G_{f_i})) \cong H_p(K_{N,2}(G_{f_i}))$ $(p > 0, i = 0, 1, ..., n)$. Consequently, the persistent barcodes of these two NC series are exactly same in positive dimensions as in Fig. 1.

Persistent spectral (PerSpect) models were proposed to study the persistence and variation of spectral information of the topological representations during a filtration process [19]. These spectral information can be used as molecular descriptors or fingerprints and combined with machine learning models for drug design. Here we study NC-bases persistent spectral models. For each NC $K_N = \{\delta_k^i; k = 0, 1, ...; i = 1, 2, ...\}$ in Eqs. (3) and (4), we can naturally assign an orientation on it. Its k-th boundary matrix B_k can be defined as follows,

$$B_k(i,j) = \begin{cases} 1, & \text{if } \delta_i^{k-1} \subset \delta_j^k \text{ and } \delta_i^{k-1} \sim \delta_j^k \\ -1, & \text{if } \delta_i^{k-1} \subset \delta_j^k \text{ and } \delta_i^{k-1} \not\sim \delta_j^k \\ 0, & \text{if } \delta_i^{k-1} \not\subset \delta_j^k. \end{cases}$$

Here $\delta_i^{k-1} \subset \delta_j^k$ means that δ_i^{k-1} is a face of δ_j^k and $\delta_i^{k-1} \not\subset \delta_j^k$ means the opposite. The notation $\delta_i^{k-1} \sim \delta_j^k$ means the two simplexes have the same orientation, i.e., oriented similarly, and $\delta_i^{k-1} \not\sim \delta_j^k$ means the opposite. Similarly, we can generate the $k + 1$-th boundary matrix B_{k+1} and the k-th Hodge (combinatorial) Laplacian matrix is defined as $L_k = B_k^T B_k + B_{k+1} B_{k+1}^T$. Further, from the Eqs. (3) and (4), two sequences of Hodge Laplacian matrixes can be generated. These matrixes characterize the interactions between protein and ligand atoms at various different scales. The spectral information derived from these Hodge Laplacian matrixes are used for the characterization of protein-ligand interactions. Note that 0-th Hodge Laplacian matrices represent topological connections between vertices, while 1-th Hodge Laplacian matrixes characterize topological connections between edges. Similarly, other-higher dimensional Hodge Laplacian matrixes can be generated for higher-dimensional simplexes. More interestingly, the multiplicity of zero-eigenvalues for the k-th dimensional Hodge Laplacian matrices is β_k, i.e., the k-th Betti number, which reflects the topological information. Additionally, information from non-zero-eigenvalues indicates "geometric" properties of the simplicial complexes.

Eigenvalues can be systematically calculated from the two sequences of Hodge Laplacian matrixes. The persistence and variation of these eigenvalues during the filtration process are defined as persistent spectral. Mathematically, a series of persistent attributes, which are statistic and combinatorial properties of eigenvalues from the sequences of Hodge Laplacian matrices, are considered. They characterize eigen spectral information during the filtration process. Here we consider 10 persistent attributes, including persistent multiplicity of zero-eigenvalues, persistent minimum value of non-zero-eigenvalues, persistent maximum value, persistent average value, persistent standard deviation, persistent sum of eigenvalues, persistent generalized mean energy, and three persistent spectral moments (with order -1, 2 and 3).

3   Experiments

The prediction of protein-ligand binding affinity is arguably the most important step in virtual screening and AI-based drug design. Here we demonstrate our NC-based machine learning models can achieve state-of-the-art results in binding affinity prediction.

3.1   Data

We consider three most commonly-used datasets from PDBbind databank, including PDB-v2007, PDB-v2013 and PDB-v2016, as benchmark for our NC-based machine learning models. The datasets can be download from http://www.pdbbind-cn.org. Following the general procedure, the core set is used as test set, and refined set (without core set) is used as train set.

3.2   Model Settings

To characterize the detailed interactions between protein and ligand at atomic level, we consider element-specific bipartite graph representations. More specifically, the binding core region of protein-ligand complex is decomposed into 36 atom combinations (such as C-C, C-O, etc.) with C, N, O, S from protein and C, N, O, S, P, F, Cl, Br, I from ligand, and the protein-ligand interactions are represented by 36 types of bipartite graphs from these atom combinations. For electrostatic interactions, H atoms are taken into consideration and a total of 50 types of atom combinations are considered.

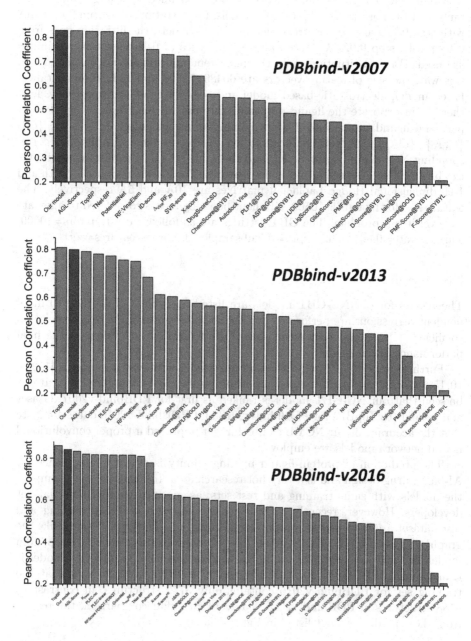

Fig. 2. The comparison of PCCs between our model and other ML (DL) models, for the prediction of protein-ligand binding affinity. The PCCs are calculated based on the core set (test set) of PDBbind-v2007, PDBbind-v2013 and PDBbind-v2016

Computationally, the binding core region is defined by using a cutoff distance of 10Å. For distance-based NC models, the filtration goes from 2Å to 10Å with step 0.1Å, and for electrostatic-based NC models, the filtration goes from 0 to 1 with step 0.02. As stated above, ten types of 0-D persistent attributes are used. For 1-D situation, only the persistent multiplicity is considered. In this way, the size of feature vectors are 60480, 52500 and 112980 for distance-based model, electrostatic-based model and combined model respectively. Further, we incorporate the ligand properties into our model. Similarly, we decompose the ligand atoms into 7 atom combinations, including $\{C\}$, $\{N\}$, $\{O\}$, $\{C, N\}$, $\{C, O\}$, $\{C, N, O\}$ and all-atoms. Based on them, graphs and corresponding NCs can be constructed, and the same set of 10 persistent attributes can be calculated as used as molecular features. The filtration values for 0-th and 1-th dimensional Hodge Laplacian matrixes are 10Å and 5Å respectively. The feature size for ligands is 10500. Gradient boosting tree is considered to alleviate the overfitting problem. The GBT settings are as follows: n_estimators=40000, learning_rate=0.001, max_depth=6, subsample=0.7, max_features=sqrt.

3.3   Results

The results for our NC-GBT models are listed in Table 1. Note that 10 independent regressions are performed and the median values of Pearson correlation coefficient (PCC) and root mean square error (RMSE) are taken as the final performance of our model.

Further, we systematically compare our model with existing models [1–3, 8, 10, 15, 16, 22, 27, 28, 32]. Detailed comparison results can be found in Fig. 2. It can be seen that our NC-GBT model is ranked as first for PDBbind-v2007 dataset and second (to TopBP) for PDBbind-v2013 and PDBbind-v2016 datasets. Note that the accuracy of our NCML can be further improved if proper convolutional neural network models are employed.

The prediction of protein-ligand binding affinity is of great importance for AI-based drug design and is a very hot research area. In Fig. 2, we only compare the models with same training and test sets, as specified by PDBbind dataset developers. However, recently some other models with different training and test dataset are proposed for protein-ligand binding affinity predictions, like the graphDelta model [12], ECIF model [25], OnionNet-2 model [30], DeepAtom model [24] and so on [2, 9, 11, 26, 29, 33, 34]. For instance, graphDelta model is trained using PDBbind-v2018, and tested in PDBbind-v2016. Note that the training set for PDBbind-v2016 contains only 3772 structures, in contrast to 8788 structures in PDBbind-v2018. To avoid an unfair comparison, the results from these models are not listed in Fig. 2.

Table 1. The PCCs and RMSEs (pK_d/pK_i) for our NC-GBT models in three test cases, i.e., PDBbind-v2007, PDBbind-v2013 and PDBbind-v2016. Four NC-GBT models are considered with features from different types of bipartite graphs. The NC-GBT(Dist) model uses features from distance-based bipartite graphs; The NC-GBT(Chrg) model uses features from electrostatic-based bipartite graphs; The NC-GBT(Dist+Chrg) model uses features from both distance-based bipartite graphs and electrostatic-based bipartite graphs; NC-GBT(Dist+Chrg+Lig) uses features from the two types of bipartite graphs and extra features from Ligands.

Dataset	Dist	Chrg	Dist+Chrg	Dist+Chrg+Lig
PDBbind-v2007	0.825(1.385)	0.815(1.422)	0.828(1.380)	0.831(1.375)
PDBbind-v2013	0.785(1.464)	0.792(1.448)	0.798(1.432)	0.799(1.426)
PDBbind-v2016	0.831(1.279)	0.833(1.279)	0.838(1.259)	0.841(1.252)
Average	0.813(1.376)	0.813(1.383)	0.821(1.357)	0.823(1.351)

4   Conclusions

In this paper, We propose neighborhood complex based molecular representation and neighborhood complex based persistent spectral model, for the first time. We reveal the deep connections between neighborhood complex and Dowker complex for bipartite graphs for the first time, and use their topological properties for molecular interaction characterization. The neighborhood complex based molecular descriptors are combined with machine learning models for drug design. It is found that our NCML model can achieve the state-of-the-art results in protein-ligand binding affinity prediction.

References

1. Afifi, K., Al-Sadek, A.F.: Improving classical scoring functions using random forest: the non-additivity of free energy terms' contributions in binding. Chem. Biol. Drug Des. **92**(2), 1429–1434 (2018)
2. Boyles, F., Deane, C.M., Morris, G.M.: Learning from the ligand: using ligand-based features to improve binding affinity prediction. Bioinformatics **36**(3), 758–764 (2020)
3. Cang, Z.X., Mu, L., Wei, G.W.: Representability of algebraic topology for biomolecules in machine learning based scoring and virtual screening. PLoS Comput. Biol. **14**(1), e1005929 (2018)
4. Cang, Z.X., Wei, G.W.: TopologyNet: topology based deep convolutional and multi-task neural networks for biomolecular property predictions. PLOS Comput. Biol. **13**(7), e1005690 (2017)
5. Chowdhury, S., Mémoli, F.: A functorial Dowker theorem and persistent homology of asymmetric networks. J. Appl. Comput. Topol. **2**(1), 115–175 (2018)
6. Dowker, C.H.: Homology groups of relations. Ann. Math., 84–95 (1952)

7. Duvenaud, D.K., et al.: Convolutional networks on graphs for learning molecular fingerprints. In: Advances in Neural Information Processing Systems, pp. 2224–2232 (2015)
8. Feinberg, E.N., et al.: PotentialNet for molecular property prediction. ACS Cent. Sci. **4**(11), 1520–1530 (2018)
9. Hassan-Harrirou, H., Zhang, C., Lemmin, T.: RoseNet: improving binding affinity prediction by leveraging molecular mechanics energies with an ensemble of 3d convolutional neural networks. J. Chem. Inf. Model. (2020)
10. Jiménez, J., Skalic, M., Martinez-Rosell, G., De Fabritiis, G.: K$_{DEEP}$: protein-ligand absolute binding affinity prediction via 3D-convolutional neural networks. J. Chem. Inf. Model. **58**(2), 287–296 (2018)
11. Jones, D., et al.: Improved protein-ligand binding affinity prediction with structure-based deep fusion inference. J. Chem. Inf. Model. **61**(4), 1583–1592 (2021)
12. Karlov, D.S., Sosnin, S., Fedorov, M.V., Popov, P.: graphDelta: MPNN scoring function for the affinity prediction of protein-ligand complexes. ACS Omega **5**(10), 5150–5159 (2020)
13. Kozlov, D.: Combinatorial Algebraic Topology, vol. 21. Springer, Heidelberg (2007). https://doi.org/10.1007/978-3-540-71962-5
14. Kozlov, D.N.: Chromatic numbers, morphism complexes, and Stiefel-Whitney characteristic classes. arXiv preprint math/0505563 (2005)
15. Li, H.J., Leung, K.S., Wong, M.H., Ballester, P.J.: Improving AutoDock Vina using random forest: the growing accuracy of binding affinity prediction by the effective exploitation of larger data sets. Mol. Inf. **34**(2–3), 115–126 (2015)
16. Liu, J., Wang, R.X.: Classification of current scoring functions. J. Chem. Inf. Model. **55**(3), 475–482 (2015)
17. Lo, Y.C., Rensi, S.E., Torng, W., Altman, R.B.: Machine learning in chemoinformatics and drug discovery. Drug Disc. Today **23**(8), 1538–1546 (2018)
18. Lovász, L.: Kneser's conjecture, chromatic number, and homotopy. J. Comb. Theory Ser. A **25**(3), 319–324 (1978)
19. Meng, Z.Y., Xia, K.L.: Persistent spectral based machine learning (PerSpect ML) for drug design. Science Advances (2021, in press)
20. Nguyen, D.D., Cang, Z.X., Wei, G.W.: A review of mathematical representations of biomolecular data. Phys. Chem. Chem. Phys. **22**, 4343–4367 (2020)
21. Nguyen, D.D., Cang, Z., Wu, K., Wang, M., Cao, Y., Wei, G.-W.: Mathematical deep learning for pose and binding affinity prediction and ranking in D3R Grand Challenges. J. Comput. Aided Mol. Des. **33**(1), 71–82 (2018). https://doi.org/10.1007/s10822-018-0146-6
22. Nguyen, D.D., Wei, G.W.: AGL-Score: algebraic graph learning score for protein-ligand binding scoring, ranking, docking, and screening. J. Chem. Inf. Model. **59**(7), 3291–3304 (2019)
23. Puzyn, T., Leszczynski, J., Cronin, M.T.: Recent Advances in QSAR Studies: Methods and Applications, vol. 8. Springer, Dordrecht (2010). https://doi.org/10.1007/978-1-4020-9783-6
24. Rezaei, M.A., Li, Y., Wu, D.O., Li, X., Li, C.: Deep learning in drug design: protein-ligand binding affinity prediction. IEEE/ACM Trans. Comput. Biol. Bioinform. (2020)
25. Sánchez-Cruz, N., Medina-Franco, J.L., Mestres, J., Barril, X.: Extended connectivity interaction features: improving binding affinity prediction through chemical description. Bioinformatics **37**(10), 1376–1382 (2021)
26. Song, T., et al.: SE-OnionNet: a convolution neural network for protein-ligand binding affinity prediction. Front. Genet. **11**, 1805 (2020)

27. Stepniewska-Dziubinska, M.M., Zielenkiewicz, P., Siedlecki, P.: Development and evaluation of a deep learning model for protein-ligand binding affinity prediction. Bioinformatics **34**(21), 3666–3674 (2018)
28. Su, M.Y., et al.: Comparative assessment of scoring functions: the CASF-2016 update. J. Chem. Inf. Model. **59**(2), 895–913 (2018)
29. Wang, K., Zhou, R., Li, Y., Li, M.: DeepDTAF: a deep learning method to predict protein-ligand binding affinity. Brief. Bioinform. (2021)
30. Wang, Z., et al.: OnionNet-2: a convolutional neural network model for predicting protein-ligand binding affinity based on residue-atom contacting shells. arXiv preprint arXiv:2103.11664 (2021)
31. Winter, R., Montanari, F., Noé, F., Clevert, D.A.: Learning continuous and data-driven molecular descriptors by translating equivalent chemical representations. Chem. Sci. **10**(6), 1692–1701 (2019)
32. Wójcikowski, M., Kukiełka, M., Stepniewska-Dziubinska, M.M., Siedlecki, P.: Development of a protein-ligand extended connectivity (PLEC) fingerprint and its application for binding affinity predictions. Bioinformatics **35**(8), 1334–1341 (2019)
33. Zhou, J., et al.: Distance-aware molecule graph attention network for drug-target binding affinity prediction. arXiv preprint arXiv:2012.09624 (2020)
34. Zhu, F., Zhang, X., Allen, J.E., Jones, D., Lightstone, F.C.: Binding affinity prediction by pairwise function based on neural network. J. Chem. Inf. Model. **60**(6), 2766–2772 (2020)

Predictive Modelling of Highly Multiplexed Tumour Tissue Images by Graph Neural Networks

Paula Martin-Gonzalez[✉], Mireia Crispin-Ortuzar, and Florian Markowetz

Cancer Research UK Cambridge Institute, University of Cambridge,
Cambridge CB2 0RE, UK
{paula.martingonzalez,mireia.crispinortuzar,
florian.markowetz}@cruk.cam.ac.uk

Abstract. The progression and treatment response of cancer largely depends on the complex tissue structure that surrounds cancer cells in a tumour, known as the tumour microenvironment (TME). Recent technical advances have led to the development of highly multiplexed imaging techniques such as Imaging Mass Cytometry (IMC), which capture the complexity of the TME by producing spatial tissue maps of dozens of proteins. Combining these multidimensional cell phenotypes with their spatial organization to predict clinically relevant information is a challenging computational task and so far no method has addressed it directly. Here, we propose and evaluate MULTIPLAI, a novel framework to predict clinical biomarkers from IMC data. The method relies on attention-based graph neural networks (GNNs) that integrate both the phenotypic and spatial dimensions of IMC images. In this proof-of-concept study we used MULTIPLAI to predict oestrogen receptor (ER) status, a key clinical variable for breast cancer patients. We trained different architectures of our framework on 240 samples and benchmarked against graph learning via graph kernels. Propagation Attribute graph kernels achieved a class-balanced accuracy of 66.18% in the development set (N = 104) while GNNs achieved a class-balanced accuracy of 90.00% on the same set when using the best combination of graph convolution and pooling layers. We further validated this architecture in internal (N = 112) and external test sets from different institutions (N = 281 and N = 350), demonstrating the generalizability of the method. Our results suggest that MULTIPLAI captures important TME features with clinical importance. This is the first application of GNNs to this type of data and opens up new opportunities for predictive modelling of highly multiplexed images.

Keywords: Graph neural networks · Highly multiplexed imaging · Imaging mass cytometry · Tumour microenvironment · Breast cancer

M. Crispin-Ortuzar and F. Markowetz—Shared senior authorship.

Electronic supplementary material The online version of this chapter (https://doi.org/10.1007/978-3-030-87444-5_10) contains supplementary material, which is available to authorized users.

M. Reyes et al. (Eds.): iMIMIC 2021/TDA4MedicalData 2021, LNCS 12929, pp. 98–107, 2021.
https://doi.org/10.1007/978-3-030-87444-5_10

1   Introduction

Cancer cells in a tumour are embedded in complex tissues including infiltrating immune and inflammatory cells, stromal cells, and blood vessels. The collection of these diverse cells is known as the tumour microenvironment (TME) [24]. Several studies have highlighted the clinical importance of tissue architecture and cell-cell interactions in the TME [12], possibly explained by different selective pressures they exert on tumour evolution [22]. Capturing spatial organization of the TME and its impact on patients is thus a critical problem in cancer medicine.

The study of the TME has been boosted by the development of highly multiplexed imaging technologies, which measure a large number of markers per tissue section and capture both the complex cellular phenotypes in the TME and their spatial relationships [7]. One example of highly multiplexed imaging is Imaging Mass Cytometry (IMC), which enables imaging at cellular resolution by labelling tissue sections with metal-isotope-tagged antibodies [5] (Fig. 1a). These novel imaging technologies have resulted in many data analysis challenges, including the segmentation of cell nuclei and quantification of proteins [20, 21] and the phenotypic clustering of cells [1]. IMC technologies are commercially available and start being used routinely world-wide, which increases the amount of available data, but also its technical variability, and new strategies for data standardization will be needed.

A natural approach to rigorously capture the information provided by IMC is graph representation learning. Graph kernels have been developed for similar problems [2, 15], and recent developments in Geometric Deep Learning via Graph Neural Networks (GNNs) have shown how to analyse interactions and topology in a data-driven manner [3, 4].

Here we propose using graph representation learning via GNNs to study IMC phenotypes together with spatial cell-to-cell interactions. We describe and evaluate MULTIPLAI (MULTIplexed PathoLogy AI), a framework that applies GNNs on highly multiplexed images for data-driven exploration of patterns in the TME. In this proof-of-concept study we used MULTIPLAI to predict Oestrogen Receptor (ER) status, a key biomarker that determines treatment options for breast cancer patients [18]. ER-positive and ER-negative breast cancers are known to be different at the histopathological, molecular, and clinical level [6, 19], which makes ER-subtyping an excellent first case study to explore and benchmark the ability of MULTIPLAI to simultaneously capture the phenotypic and spatial structure of the tumour tissue. To evaluate our method, we trained different architectures for MULTIPLAI (Fig. 1a) and benchmarked them against the Propagation Attribute Graph Kernel [17]. Then, we applied the best MULTIPLAI model to internal and external test sets (Fig. 1b) and compared the effects of different data standardization strategies.

To our knowledge, MULTIPLAI is the first framework that applies GNNs to highly multiplexed tumour tissue images. The method captures spatial interactions between complex cell phenotypes and their relationship with clinically-relevant biomarkers, opening the door to a new understanding of the dynamics and mechanisms of the TME.

2   Method

MULTIPLAI combines two key steps: the construction of a graph from imaging data, and the application of GNNs to predict clinical variables from these graphs.

2.1   Building Graphs from Highly Multiplexed Imaging

We represent an IMC image as an undirected graph $G = (V, E)$ consisting of sets of nodes V and edges $E \subseteq V \times V$. Each node corresponds to a feature vector $x_v \in \mathbb{R}^m$ where m is the number of features. The data sets we used were already segmented and we defined each cell to be a node in V. We defined the node features x_v as the mean expression of each protein analysed. The edge set E is built by connecting cells whose segmentation centroids are within 40 μm from each other. This number is approximately twice the size of a regular cell [8] as we assume that cells have to be within reach to each other to interact [12].

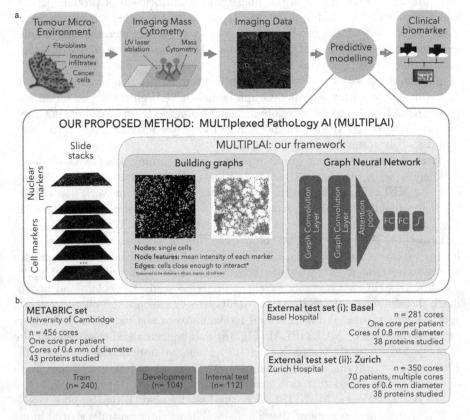

Fig. 1. (a) Overview of the proposed method. MULTIPLAI takes IMC imaging data as input for a GNN model to predict clinical biomarkers (b) Training and independent test sets used for the analysis.

2.2   Graph Neural Networks

GNNs are a type of neural network operating on a graph structure. They use an iterative procedure based on message passing from nodes to their neighbours [25]. GNNs used for graph-level binary classification contain two characteristic building blocks, graph convolution layers and graph pooling layers, that can also be combined with fully connected layers. The MULTIPLAI framework consists of the following graph convolution and pooling layers:

Graph Convolution Layers. Graph convolution layers are where the iterative procedure takes place [13]. The goal is to build an encoder that maps nodes to d-dimensional embeddings. Such embeddings are built using message propagation in local neighborhoods. On each hidden layer, message passing takes place and each node feature vector is updated considering information from its neighbours:

$$a_v^k = AGGREGATE^k(\{h_u^{k-1} : u \in N(v)\}) \tag{1}$$

$$and$$

$$h_v^k = COMBINE^k(h_v^{k-1}, a_v^k) \tag{2}$$

where $N(v)$ is the neighborhood of each node v, $h_u^k - 1$ represents the feature vector of each node in layer $k - 1$ and h_v^k refers to the updated feature vector for node v after layer k.

Graph Pooling Layers. After the neighborhood information has been propagated through the graph, the resulting node embeddings need to be combined to generate graph-level features h_g necessary for graph classification. For that, the updated node feature vectors have to be combined as follows:

$$h_g = READOUT(\{h_v^L | v \in G\}) \tag{3}$$

Where h_v^L is the updated feature vector of each node v after the last layer L.

We can combine node features using the average or the sum of the features as a readout. Another way to combine updated node features is to incorporate attention. Attention mechanisms [23] allow the network to focus on parts that are more relevant to the task being learnt. In the context of graph pooling, we can allow the network to decide the amount of focus given to each of the nodes in the graph as proposed in [16]:

$$h_g = \sum_{v=1}^{|V|} \alpha_v \cdot h_v^L \tag{4}$$

where h_g is the graph-level feature vector resulting from the combination of updated node features, $|V|$ is the number of nodes in the graph and α_v weights the attention given to each node.

Attention is calculated by feeding the updated node vectors into a parallel neural network to learn the weight given to each node through gradient descent. Thus, each α_v from Eq. 4 is calculated as:

$$\alpha_v = \text{softmax}(f_{\text{gate}}(h_v^L)) \tag{5}$$

where the softmax function ensures $\alpha_v \in (0,1)$ and $\sum_V \alpha_v = 1$, f_{gate} is the neural network used for learning the attentions and h_v^L are the node vectors.

Benchmarking: There are no state of the art methods are available to perform spatial analysis of highly multiplexed tissue data in a data driven manner to benchmark the performance of GNNs against. Therefore, we will use as a baseline the graph representation approach that uses previously defined Graph Kernels handling graphs with continuous node features. In particular, we will use the Propagation Attribute Kernel [17].

3   Data Sets

The data we used for the train, development and internal test sets is publicly available and has been previously described [1]. Briefly, this cohort consisted of 456 patients diagnosed with primary invasive breast carcinoma in Cambridge, UK. Cores of 0.6 mm in diameter were acquired and tissue micro-arrays were constructed and stained with a panel of 43 antibodies conjugated with metal tags. The slides were then analysed using the Hyperion Imaging Mass Cytometer (Fluidigm). We used the cell segmentations and marker quantification provided by the original publication [1]. The clinical information of these patients, including ER status, is also publicly available. We randomly split these cores into three data sets: training (n = 240), development (n = 104) and internal test sets (n = 112).

The data we used as the external test sets is also publicly available [9]. The first external test set contained 281 patients from the University Hospital of Basel. Pathologists reported clinical information and supervised the tissue micro-array construction of cores of 0.8 mm. The second external test set contained 70 patients from University Hospital of Zurich Hospital. Patients had multiple cores of 0.6 mm acquired and the data set contains a total of 350 slides. We used all the slides with their patient's ER status label. In both cases, samples were stained with a panel of 38 antibodies coupled to metal tags. An Hyperion Imaging System (Fluidigm) was used for acquisition. The segmentation of cells and quantification of markers again were provided by the original publication [9]. Clinical features of all cohorts are compared in Supplementary Figure 1.

4   Experiments and Results

The aim of our proof-of-concept study is to evaluate the potential of Graph Representation learning anf GNNs on IMC. Here, we focus on the well established and clinically important TME differences between ER-positive (ER+) and ER-negative (ER-) breast cancers and study how well it can be captured by MULTIPLAI, our proposed GNN-based framework. The code can be accessed here: https://github.com/markowetzlab/MULTIPLAI.

4.1   Marker Selection and Preprocessing

We only considered markers present in all the three cohorts (Metabric, Basel and Zurich). Since our goal is to quantify differences in the TME between ER+ and ER- patients, we excluded the ER marker itself as well as all other markers with highly correlated expression at the cellular level. In this way, our model focusses on the cellular composition and spatial distribution of the microenvironment and is not confounded by individual proteins. Requiring a Spearman ρ greater than 0.65 and a p-value smaller than 0.05 after Bonferroni correction resulted in the exclusion of HER2 ($\rho = 0.670$), Progesterone Receptor ($\rho = 0.739$), GATA3 ($\rho = 0.803$), E-Cadherin ($\rho = 0.699$) and PARP/Casp 3 ($\rho = 0.671$). After the correlation removal step we ended up with a set of 19 proteins from the immune system, extracellular matrix, cytokeratins and cell processes to be used (Supplementary Materials, Table 1). We used the logarithmic transform of intensities for the analysis, setting null intensities to 10^{-9}. We used the training set to define the standard scaling parameters for each marker in the single cells, and scaled the development set accordingly. In addition, we evaluated different scaling strategies to account for the different image acquisition protocols followed in each of the test datasets (Sect. 4.3).

4.2   Finding the Best Architecture

Table 1. Models tried for predicting ER status from IMC images. PA stands for Propagation Attribute. GNN stands for Graph Neural Network. Binary Cross Entropy (BCE) loss and Class Balanced (CB) accuracy reported are from the development set.

Model	Architecture		Best model	
	Number of layers	Pooling	BCE loss	CB accuracy
PA graph kernel	N.A.		7.306	66.18
GNN	1	Mean	0.174	78.4
		Sum	0.353	85.72
		Attention	0.199	86.86
	2	Mean	0.162	86.00
		Sum	0.278	81.16
		Attention	**0.159**	**90.00**

As a baseline method to check the potential of graph representation in this problem, we decided to use the Propagation Attribute graph kernel. This model was implemented in Grakel (version 0.1b7) and sklearn (version 0.22.1). The preserved distance metric on local sensitive hashing was L1 normalization. We performed a grid search of the bin width parameter (1, 3, 7, 9 and 15) selecting the model that resulted in lower BCE loss.

For MULTIPLAI, we designed six GNN architectures with different numbers of graph convolution layers and different pooling methods (Sect. 2.2). A graphical description is shown in Supplementary Figure 2. A single fully connected layer served as f_{gate} for the attention pooling. Graph level features obtained after pooling in all cases were fed into two fully connected layers and a sigmoid function, whose outputs were rounded to obtain binary predictions. We used Binary Cross Entropy (BCE) Loss and Adam optimizer. Architectures were implemented using Deep Graph Library (version 0.4.3) and Pytorch (version 1.5.1).

We trained each architecture on the training set using mini-batches, and for each epoch we evaluated the loss and class-balanced accuracy in the development set. We implemented early stopping by finding the minimum development set loss and allowing 20 epochs of patience before stopping. We performed a grid search for hyper-parameter optimization: batch size of 10, 20 and 30, hidden dimensions of fully connected layers 50, 100 and 150, dropout of 0.3 and 0.5, learning rate of 0.01 and 0.001, weight decay of 0.1, 0.01 and 0.01 and graph hidden dimensions of 50, 100 and 150 in the cases with two graph convolutions. In each case, we optimised each hyperparameter to give the a lower loss and lower standard deviation in the development set (Table 1).

We found some signal showing in the graph kernel model (Class Balanced Accuracy of 66.18) and, when moving to GNNs, even shallow architectures were able to capture the TME differences in ER+ and ER− patients with the lowest balanced accuracy being 78.40. This suggests the potential of graph representation learning for this problem and overall, GNN models do significantly better than the Graph Kernel. Additionally, the losses in the models with two layers (0.162, 0.278 and 0.159) are significantly lower than those with only one (0.174, 0.353 and 0.199). We selected as the best model the combination of two graph convolutions and attention pooling for the next sections as it reached the lower binary cross entropy loss (0.159) on the development set.

4.3   Comparing Standardization Strategies

We then applied the best model identified in the last section to the internal and two external test cohorts (Fig. 1b). This step is complicated by systematic biases in the data: The three cohorts come from three different institutions, use different sizes of tissue, use different numbers of markers and show differences in the global staining distribution across slides. To alleviate these biases, we compared two different stain standardization strategies.

Strategy 1: Fixed Scaling. The best architecture identified in Sect. 4.2 was applied to the internal and external test cohorts (Table 2, Supplementary Table 2), using the same fixed standard scaling applied on the development set. A balanced accuracy of 78.96 is seen in the internal test cohort but performance declines in the Zurich and particularly the Basel set (balanced accuracies of 68.19 and 60.16 respectively). An analysis of the stain histograms (Supplementary materials, Figure 3a) reveals that the two external sets are systematically different. Since IMC has been used so far only on same-institution descriptive studies, the field has not yet come across the need to define standardization strategies to enable inter-institution comparisons. To correct for the biases, we compared an additional normalization strategies.

Strategy 2: Independent Scaling. To explore whether we could correct for the staining differences, we scaled the intensity distributions for each marker on each dataset independently (Supplementary materials, Figure 3b). When validating the initial model from Sect. 4.3 using this version of scaling, the performance on the external test sets improves significantly (from a balanced accuracy of 60.16 to 65.71 in the Basel set and from 68.19 to 75.10 in the Zurich set) (Table 2, Supplementary Table 2). Importantly, the performance degradation observed in the external sets in all strategies reflected known experimental differences. The Basel dataset, in particular, had a bigger core size (0.8 mm) than the rest of the sets (0.6 mm), resulting in significantly larger graphs than the other two.

Table 2. Metrics for each of the test cohorts and preprocessing conditions

Class balanced accuracy		
	Strategy 1	Strategy 2
Internal test (Metabric)	78.96	**78.96**
External test (Basel)	60.16	**65.71**
External test (Zurich)	68.19	**75.10**

5 Discussion

We have presented a proof-of-concept study showing how GNNs can be applied to IMC data for predictive modelling of clinical biomarkers. We showed that our proposed framework, called MULTIPLAI, was able to capture TME differences between ER+ and ER− patients. These results signal a promising new avenue for the computational analysis of highly multiplexed images.

Some features of the method can be further optimized: First, we pragmatically fixed the distance threshold for building the graph and we plan to perform an ablation study of this distance to check the stability of the graph construction step.

Second, the influence of scaling and graph size to model performance suggests that IMC data standarization is a key challenge to develop methods that are robust across institutions. Data augmentation could help increasing the robustness of the MULTIPLAI framework to these changes between institutions but there is little literature available on data augmentation for graphs [14,26]. This suggests that the need to propose such methods is also an open challenge in the GNN field. The descriptive work we presented in Sect. 4.2 could be used as a benchmark for new data augmentation techniques.

Third, it will be important to address the explainability of the framework in future work. Exploring the node and feature importance driving different predictions will help cancer biologists understand that the method is capturing known TME differences as well as discover new patterns not yet reported. Unlike in the Convolutional Neural Networks domain, very few alternatives are available to study the explainability of decisions in GNNs [10,11], and more work in the area is needed.

Finally, we compared the suitability of two graph learning approaches (GNNs and Graph Kernels) in this study but it would be interesting to compare the performance of graph learning to Convolutional Neural Networks (CNNs) in this setting. Although off-the-shelf CNNs are built for RGB images, the input tensor could be extended to match the number of channels in IMC images.

To conclude, this proof-of-concept study shows that the analysis of highly multiplexed images in cancer can benefit from using GNN-based methods. These frameworks spatially integrate the different phenotypes of the TME that could be used to perform data-driven clinical decisions.

References

1. Ali, H.R., et al.: Imaging mass cytometry and multiplatform genomics define the phenogenomic landscape of breast cancer. Nat. Cancer **1**(2), 163–175 (2020). https://doi.org/10.1038/s43018-020-0026-6

2. Borgwardt, K.M., Ghisu, M.E., Llinares-López, F., O'Bray, L., Rieck, B.: Graph kernels: state-of-the-art and future challenges. CoRR abs/2011.0 (2020). https://arxiv.org/abs/2011.03854

3. Bronstein, M.M., Bruna, J., Lecun, Y., Szlam, A., Vandergheynst, P.: Geometric deep learning: going beyond Euclidean data, July 2017. https://doi.org/10.1109/MSP.2017.2693418

4. Cao, W., Yan, Z., He, Z., He, Z.: A comprehensive survey on geometric deep learning. IEEE Access **8**, 35929–35949 (2020). https://doi.org/10.1109/ACCESS.2020.2975067

5. Chang, Q., Ornatsky, O.I., Siddiqui, I., Loboda, A., Baranov, V.I., Hedley, D.W.: Imaging mass cytometry. Cytometry Part A **91**(2), 160–169 (2017). https://doi.org/10.1002/cyto.a.23053. http://doi.wiley.com/10.1002/cyto.a.23053

6. Curtis, C., et al.: The genomic and transcriptomic architecture of 2,000 breast tumours reveals novel subgroups. Nature **486**(7403), 346–352 (2012). https://doi.org/10.1038/nature10983. https://www.nature.com/articles/nature10983

7. Giesen, C., et al.: Highly multiplexed imaging of tumor tissues with subcellular resolution by mass cytometry. Nat. Methods **11**(4), 417–422 (2014). https://doi.org/10.1038/nmeth.2869. https://www.nature.com/articles/nmeth.2869

8. Hao, S.J., Wan, Y., Xia, Y.Q., Zou, X., Zheng, S.Y.: Size-based separation methods of circulating tumor cells, February 2018. https://doi.org/10.1016/j.addr.2018.01.002

9. Jackson, H.W., et al.: The single-cell pathology landscape of breast cancer. Nature **578**(7796), 615–620 (2020). https://doi.org/10.1038/s41586-019-1876-x

10. Jaume, G., et al.: Quantifying explainers of graph neural networks in computational pathology. Technical report (2021)

11. Jaume, G., et al.: Towards explainable graph representations in digital pathology. Technical report (2020)
12. Kamińska, K., et al.: The role of the cell-cell interactions in cancer progression. J. Cell. Mol. Medi. **19**(2), 283–296 (2015). https://doi.org/10.1111/jcmm.12408
13. Kipf, T.N., Welling, M.: Semi-supervised classification with graph convolutional networks. In: 5th International Conference on Learning Representations, ICLR 2017 - Conference Track Proceedings, September 2016. http://arxiv.org/abs/1609.02907
14. Kong, K., et al.: FLAG: adversarial data augmentation for graph neural networks, October 2020. http://arxiv.org/abs/2010.09891
15. Kriege, N.M., Johansson, F.D., Morris, C.: A survey on graph kernels. Appl. Netw. Sci. **5**(1), 6 (2020). https://doi.org/10.1007/s41109-019-0195-3
16. Li, Y., Tarlow, D., Brockschmidt, M., Zemel, R.: Gated graph sequence neural networks. In: 4th International Conference on Learning Representations, ICLR 2016 - Conference Track Proceedings, November 2015. http://arxiv.org/abs/1511.05493
17. Neumann, M., Garnett, R., Bauckhage, C., Kersting, K.: Propagation kernels: efficient graph kernels from propagated information. Mach. Learn. **102**(2), 209–245 (2016). https://doi.org/10.1007/s10994-015-5517-9
18. Onitilo, A.A., Engel, J.M., Greenlee, R.T., Mukesh, B.N.: Breast cancer subtypes based on ER/PR and Her2 expression: comparison of clinicopathologic features and survival. Clin. Med. Res. **7**(1–2), 4–13 (2009). https://doi.org/10.3121/cmr.2008.825. http://www.clinmedres.org/content/7/1-2/4.full
19. Putti, T.C., et al.: Estrogen receptor-negative breast carcinomas: a review of morphology and immunophenotypical analysis. Mod. Pathol. **18**(1), 26–35 (2005). https://doi.org/10.1038/modpathol.3800255
20. Schapiro, D., et al.: histoCAT: analysis of cell phenotypes and interactions in multiplex image cytometry data. Nat. Methods **14**(9), 873–876 (2017). https://doi.org/10.1038/nmeth.4391. http://www.nature.com/articles/nmeth.4391
21. Somarakis, A., Van Unen, V., Koning, F., Lelieveldt, B., Hollt, T.: ImaCytE: visual exploration of cellular micro-environments for imaging mass cytometry data. IEEE Trans. Vis. Comput. Graph. **27**(1), 98–110 (2021). https://doi.org/10.1109/TVCG.2019.2931299
22. Turajlic, S., Swanton, C.: Implications of cancer evolution for drug development, July 2017. https://doi.org/10.1038/nrd.2017.78. http://tracerx.co.uk/
23. Wang, F., Tax, D.M.J.: Survey on the attention based RNN model and its applications in computer vision, January 2016. http://arxiv.org/abs/1601.06823
24. Whiteside, T.L.: The tumor microenvironment and its role in promoting tumor growth, October 2008. https://doi.org/10.1038/onc.2008.271./pmc/articles/PMC3689267/
25. Xu, K., Hu, W., Leskovec, J., Jegelka, S.: How powerful are graph neural networks? arXiv, October 2018. http://arxiv.org/abs/1810.00826
26. Zhao, T., Liu, Y., Neves, L., Woodford, O., Jiang, M., Shah, N.: Data augmentation for graph neural networks. arXiv, July 2020. http://arxiv.org/abs/2006.06830

Statistical Modeling of Pulmonary Vasculatures with Topological Priors in CT Volumes

Yuki Saeki[1], Atsushi Saito[1], Jean Cousty[2], Yukiko Kenmochi[2]([✉]),
and Akinobu Shimizu[1]

[1] Tokyo University of Agriculture and Technology, Tokyo, Japan
[2] LIGM, Univ Gustave Eiffel, CNRS, ESIEE Paris, Marne-la-Vallée, France
yukiko.kenmochi@esiee.fr

Abstract. A statistical appearance model of blood vessels based on variational autoencoder (VAE) is well adapted to image intensity variations. However, images reconstructed with such a statistical model may have topological defects, such as loss of bifurcation and creation of undesired hole. In order to build a 3D anatomical model of blood vessels, we incorporate topological prior into the statistical modeling. Qualitative and quantitative results on 2567 real CT volume patches and on 10000 artificial ones show the efficiency of the proposed framework.

Keywords: Statistical model · Topological data analysis · Variational autoencoder · Vasculature

1 Introduction

A statistical model, which represents a variation in shape and intensity of an organ with a small number of parameters, plays a key role as priors for medical image analysis [5,8]. Many statistical shape models have been proposed for parenchymatous organs, such as the liver [1,3,14]. However, as far as we know, less study made for building a statistical appearance model that accurately represents variations of blood vessels, whose geometrical and appearance features are useful for the diagnosis of diseases, e.g. pneumonia. This is due to the fact that a simple statistical model is not suitable to describe diverse variations of vessels including difference in not only direction and diameter but also topology such as branching. For example, Principal Component Analysis (PCA), which is a classical statistical analysis method, cannot deal with such complicated distribution. Recently proposed methods, such as manifold learning and deep learning [2], might solve the problem in the modeling of blood vessels. Along this line, the approach employing a Variational AutoEncoder (VAE) [12] and its extension

Partially supported by MEXT/JSPS KAKENHI Grant Number JP18H03255 and CNRS International Emerging Action Project DiTopAM.

M. Reyes et al. (Eds.): iMIMIC 2021/TDA4MedicalData 2021, LNCS 12929, pp. 108–118, 2021.
https://doi.org/10.1007/978-3-030-87444-5_11

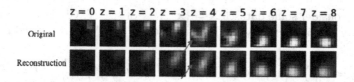

Fig. 1. Cross sections of an original volume patch of $9 \times 9 \times 9$ [voxel] of vasculatures (top) and its failed reconstruction (missing bifurcation marked by the red arrow) by a statistical model based on β-VAE [13] (bottom). (Color figure online)

β-Variational AutoEncoder (β-VAE) [9] was previously proposed to build a statistical model of vasculatures of lung CT volumes [13]. Nevertheless, it has been pointed out that the images generated by the statistical model suffer from topological artifacts, such as disconnection of vessels, undesired holes, false cycles, and missing bifurcations (see Fig. 1).

This article proposes a method for building a statistical intensity model of pulmonary vasculatures in CT volume patches that incorporates topological prior of vasculatures in the statistical modeling, which is an incremental contribution over the previous work [13]. We use persistent homology [6] to give topological constraints in neural network [4], which result in topologically correct representation of vasculatures. In particular, in the case of bifurcations and undesired hole artifacts, qualitative and quantitative evaluations show that taking topological prior into account during learning indeed allows for improving the topological soundness of reconstructed images while keeping a satisfying appearance level. We also investigate its combination with a multi-scale approach based on mathematical morphology [15] when the topological correction (hole closing or vessel re-connection) forced by the topological loss is too "thin".

2   Appearance Model of 3D Blood Vessels Based on β-VAE

Statistical variations of vessels in chest CT volume patches are modeled in two steps: modeling appearance of vessels using a conventional network, or β-VAE, as a baseline model and incorporating topological prior to the model. In this section, we explain the first step.

Vessel appearance in a volume patch is modeled by the β-VAE (Fig. 2) [13] that consists of four layers, or two fully connected layers with 200 units followed by a rectified linear unit (ReLU), a latent layer, and a fully connected layer with 729 units followed by a sigmoid function to generate an output vector. Note that a volume patch of $9 \times 9 \times 9$ [voxel] is transformed into a vector of size 729 for the input and the output vector is reconstructed to a volume patch.

Let us consider some dataset $\mathbf{X} = \{\mathbf{x}^{(i)}\}_{i=1}^N$ of size N. Given a volume patch $\mathbf{x}^{(i)}$ and its reconstruction $\mathbf{y}^{(i)}$, the loss function of β-VAE is defined by

$$\mathcal{L}_{\beta\text{-VAE}} = -\frac{\beta}{2} \sum_{j=1}^{J} \left(1 + \log((\sigma_j^{(i)})^2) - (\mu_j^{(i)})^2 - (\sigma_j^{(i)})^2\right) + \frac{1}{2}\|\mathbf{x}^{(i)} - \mathbf{y}^{(i)}\|_2^2 \quad (1)$$

where $\mu^{(i)}$ and $\sigma^{(i)}$ are the mean and variance of the latent space variables $\mathbf{z}^{(i)}$, both of which are computed using the training data and $\mathbf{x}^{(i)}$. Note that j represents the dimensional index of $\mathbf{z}^{(i)}$, and J is the dimension of the latent space. The first term on the right-hand side is the Kullback-Leibler distance between the predicted distribution in the latent space and the Gaussian distribution $\mathcal{N}(0, I)$, while the second term is the reconstruction error of $\mathbf{x}^{(i)}$. The weight β determines a balance between the two terms.

This statistical model allows us to describe vessels using a small number of parameters and to generate vessels by changing latent parameters. The encoder of the trained β-VAE has the former function, while the decoder has the latter function. Thanks to these two functions, this model enables us to compress a given vessel and also to generate a vessel.

3   Topological Loss Function Based on Persistent Homology

The second step of the statistical modeling is to incorporate topological prior into the appearance model of Sect. 2. To this end, computational topology [6] is used to fine-tune the network. We focus on persistent homology [17] and incorporate the topological loss function [4] in the fine-tuning phase.

3.1   Persistent Homology

Persistent Homology (PH) aims at describing the topological features of data [6]. Given a function or image defined over a simplicial complex, PH represents the evolution of the topological features of the lower level sets (also called threshold sets) of this function. Thus, the image resulting from the previously presented network must be first converted into a weighted simplicial complex before computing PH. In a complex of dimension d, each topological feature has a dimension

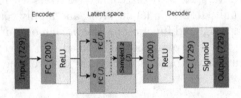

Fig. 2. Network architecture of β-VAE [13] for statistical modeling of vessels

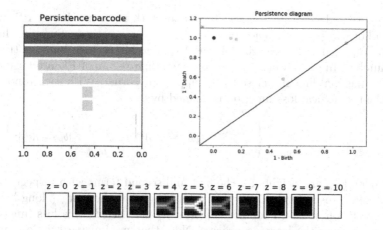

Fig. 3. Persistence barcode (top, left) and diagram (top, right) of a 3D bifurcated vessel image with a frame padded (bottom) (Color figure online)

between 0 and $d-1$ and, for each dimension k, the number of topological features of dimension k is known as the k-th Betti number, denoted by β_k. In a 3D space, the features of dimension 0 (resp. 1, and 2) correspond to the connected components (resp. to the tunnels, and to the cavities) of the complex. When browsing the successive threshold sets of a weighted complex, at each level, it is possible to track the new topological features that appear, those that disappear at this level, and those that persist from the previous level. Therefore, for each topological feature, it is possible to compute the level at which it appears and the one at which it disappears. These levels are called the birth and death times of the topological feature, respectively. PH then produces a *Persistence Barcode (PB)* such as the one shown in Fig. 3, where a bar is displayed for each topological feature and where the starting and ending points of the bar correspond to the birth and death times of the considered feature. In the figure, blue (resp. green and red) bars represent the features of dimension 0 (resp. 1 and 2) composing the Betti number β_0 (resp. β_1 and β_2). A *Persistence Diagram (PD)* can also be drawn where each feature is represented as a point whose coordinates correspond to its birth and death times. The length of a bar in a PB indicates the lifetime of each topological feature. When the lifetime is long, the corresponding point of a PD is plotted close to the upper left corner and considered as a stable feature. In this example, blue and red points are at stable positions. When the lifetime is short, the corresponding points are plotted close to the diagonal line and such components are considered as noise.

3.2   Topological Loss Functions and Priors

Based on the idea presented in [4], we assume that topological information of 3D blood vessels is known a priori, and incorporate this knowledge into deep learning based modeling.

Let us consider the l-th longest bar of the k-th Betti number of PB, whose birth and death times are denoted by $b_{k,l}$ and $d_{k,l}$, respectively. Suppose that we have the topological prior such that the PB contains β_k^* long bars for the k-th Betti number. In other words, those corresponding points of PD are located far from the diagonal line and close to the upper left corner. With such topological prior, the topological loss function is defined by

$$\mathcal{L}_{\text{topo}}(\beta_0^*, \beta_1^*, \beta_2^*) = \sum_{k \in \{0,1,2\}} \left\{ \lambda_k^* \sum_{l=1}^{\beta_k^*} (1 - |b_{k,l} - d_{k,l}|^2) + \lambda_k \sum_{l=\beta_k^*+1}^{\infty} |b_{k,l} - d_{k,l}|^2 \right\} \quad (2)$$

where λ_k^* and λ_k are weights for the first and second terms with respect to k-th Betti number, respectively. Note that we assume that the first β_k^* long bars are the "correct" ones. Given topological prior β_k^* for each k, this loss function is minimized during the learning process. Note that minimizing this loss implies increasing the lengths of the β_k^* long bars of each k-th Betti number on PB (the first term of (2)) and decreasing the lengths of the rest (the second term of (2)). This will encourage the β_k^* points of PD in the first term to move to the upper left point while the rest moves to the diagonal line.

This article is aimed at modeling 3D blood vessels in volume patches. Toward this end, we consider various topological loss functions. First, we propose to set

- $(\beta_0^*, \beta_1^*, \beta_2^*) = (1, 0, 0)$ in order to favor one connected component

existing in a volume patch, denoted by $\mathcal{L}_{\text{topo}}(1, 0, 0)$. As this topological prior is not sufficient to distinguish a single tube and a bifurcation structure, we also consider the padded volume patches with a one-voxel border of value 1 (see Fig. 3 (bottom)), to which we propose to set

- $(\beta_0^*, \beta_1^*, \beta_2^*) = (1, 1, 1)$ for a single tube structure, and
- $(\beta_0^*, \beta_1^*, \beta_2^*) = (1, 2, 1)$ for a single bifurcation structure (Y-shaped)

as the topological priors, denoted by $\mathcal{L}_{\text{topo}}^{\square}(1, 1, 1)$ and $\mathcal{L}_{\text{topo}}^{\square}(1, 2, 1)$. A similar idea of using padded images for 2D image segmentation can be seen in [10].

In some cases, the topological correction (hole closing or vessel re-connection) forced by the topological loss is too "thin" compared to the width of the vessel (see Fig. 4–line 4). It is then desirable to also consider a topological loss applied to an erosion of the generated image (insertion of an erosion layer before computing the topological loss), leading to multi-scale topological loss functions, where the scale is given by the size of the structuring element used in the erosion. The effect of this multi-scale loss function can be seen in Fig. 4–line 5 where the topological correction is thick enough.

4   Incorporating Topological Priors into Statistical Model

The topological priors proposed in the previous section are incorporated into the deep learning generator by using the following loss function

$$\mathcal{L} = \mathcal{L}_{\beta\text{-VAE}} + \mathcal{L}_{\text{topo}} \quad (3)$$

instead of simply using $\mathcal{L}_{\beta\text{-VAE}}$ of (1). Note that the effect of homology based loss on the total loss can be controlled by setting values of weights λ_k^* and λ_k inserted in $\mathcal{L}_{\text{topo}}$ (see Eq. (2)). In this article, the function $\mathcal{L}_{\text{topo}}$ can be either the topological function that defines the number of connected components $\mathcal{L}_{\text{topo}}(1,0,0)$, the structure of non-bifurcation $\mathcal{L}_{\text{topo}}^{\square}(1,1,1)$, the structure of bifurcation $\mathcal{L}_{\text{topo}}^{\square}(1,2,1)$, their multi-scale version, or their linear combination. It should be noted that such a combination allows us to represent more complex shapes while it provides more parameter tuning and more computation time.

5   Experiments

The network is pre-trained with only the β-VAE loss (1) and fine-tuned by adding the topological loss (2). The idea behind this strategy is that the network is first optimized coarsely with the β-VAE loss and then refined with the topological loss. This is because the topological loss is likely to provide many local optimums everywhere in the search space. The method is evaluated in the following two contexts: blood vessels containing (1) hole artifacts, created in image acquisition process such as random noise, and (2) bifurcations. The topological priors are chosen depending on the contexts. Volume patches of size $9 \times 9 \times 9$ are treated and intensities of every patch are normalized to $[0,1]$. Both in the pre-training and the fine-tuning, the number of epochs is decided such that the validation loss becomes minimum.

5.1   Blood Vessels Containing Hole Artifacts

Datasets. Artificial images of blood vessels containing hole artifacts are generated from spatial lines with a Gaussian intensity profile whose center is positioned on the lines and whose parameter values are tuned (no hole image). For each image, a hole is then generated by reversing the intensity values of a ball, whose center is on the centerline of a blood vessel (hole images) and which is located inside the blood vessel. Among the generated images, 6000, 2000 and 2000 volume patches are used for training, validation and testing.

Chosen Topological Loss Function and Parameter Setting. The chosen topological prior is of one connected component $\mathcal{L}_{\text{topo}}(1,0,0)$. The latent dimensions J of β-VAE was set to 6, and the parameters of the loss function were set to $\beta = 0.1$ and $\lambda_k^* = \lambda_k = 60000$. Adam [11] was used with $\alpha = 10^{-5}, \beta_1 = 0.9, \beta_2 = 0.999$ as the optimization method, where the batch size was set to 128. The number of epochs was set to 3072 (resp. 371) for the pre-training (resp. the fine-training).

Evaluation Strategies. The performance of the statistical model for intensity distributions was evaluated by the following two indices, generalization, which measures the ability to represent unknown data correctly, and specificity, which

Table 1. Evaluations of the three methods (Pre, Fine and Fine with erosion) on their reconstructions of a volume containing a hole artifact (see Fig. 4) with three measures based on generalization, specificity and topological loss function.

	Generalization	Specificity	Topology
Pre	**0.00251**	**0.0134**	0.430
Fine	0.00478	0.0143	0.0976
Fine + Erosion	0.00728	0.0156	**0.00139**

measures the ability to eliminate an unnatural shape [16]; they are made by comparison with test images and their reconstructed images. Note that the smaller index leads to better performance since they measure the distances between the image generated by the model and the original images.

For evaluating the topological performance of the model, the topological loss function was calculated with $\lambda_k^* = \lambda_k = 1$ (the smaller the index, the better the performance). The null hypothesis that the values achieved by each method (Pre, Fine and Fine (Erosion)) have the same distribution was tested: Wilcoxon's signed quartiles test was used for generalization and topological loss function, and Mann-Whitney's U test was used for specificity.

Results. The method without topological prior is denoted as Pre while the proposed one with topological prior without/with multiscale is denoted as Fine and Fine (Erosion), respectively. Table 1 shows that Fine is inferior to Pre in the evaluation of intensity values, but it improves the performance in the evaluation of topology. However, the 4th row of Fig. 4 shows that the result still contains an artifact. With the multiscale version of this topological loss, this is improved qualitatively (the 5th row of Fig. 4) and quantitatively with topological evaluation (Table 1). Null hypothesis (H0: there is no difference in performance between any pair of the three methods, Pre, Fine and Fine (Erosion))

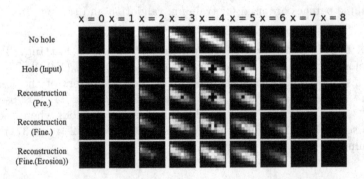

Fig. 4. Modeling of a volume containing a hole artifact (Input): resulting reconstruction without/with topological prior (Pre/Fine), and with the multiscale version (Erosion).

Table 2. Evaluations of the two methods (Pre and Fine) on their reconstructions of a volume containing a bifurcation (see Fig. 5) with three measures based on generalization, specificity and bottleneck distance with the original image via the PDs.

	Generalization	Specificity	Bottleneck distance
Pre	**0.00704**	**0.0283**	0.288
Fine	0.0102	0.0410	**0.236**

was rejected ($p < 0.01$) for all evaluation measures, except for the specificity between Fine and Fine (Erosion).

5.2 Blood Vessels with Bifurcations

Datasets. CT images taken at Tokushima University Hospital (TUAT Ethics Committee Application No. 30-29) were used in the experiments. From 47 cases, volume patches of size $9 \times 9 \times 9$ containing Y-shaped blood vessels with a thickness of approximately from 1 to 4 mm, located between the hilar and peripheral regions, were extracted using a method based on Hessian filter [7]. Among them, 1533, 517 and 517 volume patches are used for training, validation and testing. The training images were augmented by flipping the original volume patches with probability 0.5 before inputting to the network.

Chosen Topological Loss Function and Parameter Setting. The network was pre-trained with only the β-VAE loss (1) and fine-tuned by adding topological prior (2) of bifurcation case $\mathcal{L}^{\square}_{\text{topo}}(1, 2, 1)$. The latent dimensions J of β-VAE was set to 22, which corresponds to the number of axes of PCA whose cumulative contribution exceeds 0.8 using the training and validation data. The parameters of the loss function were experimentally set to $\beta = 0.1$ and $\lambda_k^* = \lambda_k = 50$. Adam [11] was used with $\alpha = 10^{-3}, \beta_1 = 0.9, \beta_2 = 0.999$ as the optimization method, where the batch size was set to 128. To reduce the computational cost, the topological loss was calculated using $\frac{1}{10}$ of the batch size. The number of epochs was set to 346 (resp. 36) for the pre-training (resp. the fine-training).

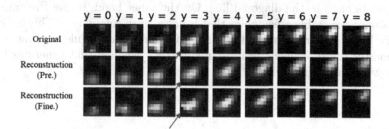

Fig. 5. Modeling of a volume containing a bifurcation (Original): resulting reconstruction without/with topological prior (Pre/Fine)

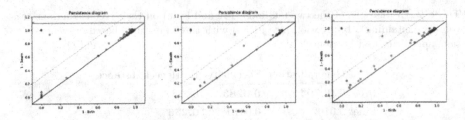

Fig. 6. PDs of the original (left) and reconstructed images of Fig. 5: Pre. (middle) and Fine (right).

Evaluation Strategies. Similarly to the previous experiment, generalization and specificity were calculated. For evaluating the topological performance of the model, the Bottleneck distance [6] between the PDs of the original and reconstructed images was used rather than the topological loss function, which is simpler to be calculated, as we consider in this experiment that the reconstruction must be close to the original. Note that the smaller the distance, the better the performance. Wilcoxon's signed quartiles test was used for Bottleneck distance as well as generalization while Mann-Whitney's U test was used for specificity.

Results. The method without topological prior is denoted as Pre and the proposed method with topological prior is denoted as Fine. Table 2 shows that the Fine is inferior to the Pre in the evaluation of intensity values, but it improves the performance in the evaluation of topology. Null hypothesis (H0: there is no difference in performance between Pre and Fine) was rejected ($p < 0.01$) for all evaluation measures.

The reconstructed images are shown in Fig. 5 with the red arrows which indicate the typical difference between the two methods. The arrows confirm that the proposed model (Fine) improves the representation of bifurcation compared to the conventional model (Pre).

The PDs corresponding to the original and reconstructed images in Fig. 5 are shown in Fig. 6. In the original (left) and the Fine (right), two green points are positioned close to the left corner point $(0, 1)$ and apart from other noise components close to the diagonal line. On the other hand, in the Pre (middle) there is only one stable green point that is not a noise component. This indicates that the Pre cannot represent bifurcations of blood vessels, while the Fine, which incorporates the topological prior, can construct a model that generates blood vessels with bifurcations closer to the ones of the original.

6   Conclusion

In this article, we tackled the problem of modeling blood vessels with topological correctness based on a statistical appearance model, which has been a challenge for conventional methods. In order to incorporate topological priors of bifurcations and undesired hole existence into statistical modeling, we integrated topological loss into the deep generative model (β-VAE). More precisely, the network is pre-trained with only the β-VAE loss and fine-tuned by adding the topological loss. The topological is indeed used for refining the network as it may provide local optimums everywhere in the search space. In the experiments, we used a dataset of vascular patches with bifurcation structures to build a model using the proposed method. The experimental results showed that the proposed method improved the performance of the topology evaluation values while the performance of the intensity evaluation value decreased. Qualitatively, we also identified improvements of hole closing and representation of bifurcation structures in the reconstructed volumes. Future work would be to apply the constructed statistical appearance model of blood vessels to associated tasks such as vessel segmentation and super-resolution.

References

1. Bailleul, J., Ruan, S., Bloyet, D., Romaniuk, B.: Segmentation of anatomical structures from 3D brain MRI using automatically-built statistical shape models. In: ICIP 2004, vol. 4, pp. 2741–2744 (2004)
2. Bengio, Y., Courville, A., Vincent, P.: Representation learning: a review and new perspectives. IEEE Trans. Pattern Anal. Mach. Intell. **35**(8), 1798–1828 (2013)
3. Blackall, J.M., King, A.P., Penney, G.P., Adam, A., Hawkes, D.J.: A statistical model of respiratory motion and deformation of the liver. In: MICCAI 2001, pp. 1338–1340 (2001)
4. Clough, J.R., Byrne, N., Oksuz, I., Zimmer, V.A., Schnabel, J.A., King, A.P.: A topological loss function for deep-learning based image segmentation using persistent homology. IEEE Trans. Pattern Anal. Mach. Intell. (2020)
5. Cremers, D., Rousson, M., Deriche, R.: A review of statistical approaches to level set segmentation: integrating color, texture, motion and shape. Int. J. Comput. Vision **72**(2), 195–215 (2007)
6. Edelsbrunner, H., Harer, J.: Computational Topology - An Introduction. American Mathematical Society (2010)
7. Frangi, A.F., Niessen, W.J., Vincken, K.L., Viergever, M.A.: Multiscale vessel enhancement filtering. In: MICCAI 1998, vol. 1496, pp. 130–137 (1998)
8. Heimann, T., Meinzer, H.P.: Statistical shape models for 3D medical image segmentation: a review. Med. Image Anal. **13**(4), 543–563 (2009)
9. Higgins, I., et al.: β-VAE: learning basic visual concepts with a constrained variational framework. In: 2nd ICLR (2017)
10. Hu, X., Li, F., Samaras, D., Chen, C.: Topology-preserving deep image segmentation. In: NeurIPS 2019 (2019)
11. Kingma, D.P., Ba, J.: Adam: a method for stochastic optimization. In: 3rd ICLR (2015)
12. Kingma, D.P., Welling, M.: Auto-encoding variational bayes. In: 2nd ICLR (2014)
13. Saeki, Y., Saito, A., Ueno, J., Harada, M., Shimizu, A.: Statistical intensity model of lung vessels in a CT volume using β-VAE. In: CARS (2019)
14. Saito, A., Shimizu, A., Watanabe, H., Yamamoto, S., Nawano, S., Kobatake, H.: Statistical shape model of a liver for autopsy imaging. Int. J. Comput. Assist. Radiol. Surg. **9**(2), 269–281 (2014)

15. Serra, J.: Image Analysis and Mathematical Morphology. Academic Press, USA (1983)
16. Styner, M.A., et al.: Evaluation of 3D correspondence methods for model building. In: Information Processing in Medical Imaging, vol. 2732, pp. 63–75 (2003)
17. Zomorodian, A., Carlsson, G.: Computing persistent homology. Discrete Comput. Geom. **33**(2), 249–274 (2005)

Topological Detection of Alzheimer's Disease Using Betti Curves

Ameer Saadat-Yazdi[✉], Rayna Andreeva, and Rik Sarkar

School of Informatics, University of Edinburgh, Edinburgh, UK
{s1707343,r.andreeva}@sms.ed.ac.uk, rsarkar@inf.ed.ac.uk

Abstract. Alzheimer's disease is a debilitating disease in the elderly, and is an increasing burden to the society due to an aging population. In this paper, we apply topological data analysis to structural MRI scans of the brain, and show that topological invariants make accurate predictors for Alzheimer's. Using the construct of Betti Curves, we first show that topology is a good predictor of Age. Then we develop an approach to factor out the topological signature of age from Betti curves, and thus obtain accurate detection of Alzheimer's disease. Experimental results show that topological features used with standard classifiers perform comparably to recently developed convolutional neural networks. These results imply that topology is a major aspect of structural changes due to aging and Alzheimer's. We expect this relation will generate further insights for both early detection and better understanding of the disease.

Keywords: Topological data analysis · Alzheimer's disease · MRI

1 Introduction

Alzheimer's disease (which is often abbreviated as AD) and other neurodegenerative diseases are closely associated with aging. As average life expectancy increases worldwide, the number of patients with brain aging and associated diseases will rise rapidly. The current estimated number of Alzheimer's patients is around 47 million, which is projected to increase to 152 million by year 2050 [25].

Deterioration of the brain manifests as several complex structural, chemical and functional changes, making it challenging to distinguish diseases from aging. For example, with age, cerebral ventricles expand and cortical thickness decreases; lesions and atrophies arise [13]; the brain volume itself contracts with old age, while gray and white matter volumes are known to expand and contract at different times in the life cycle. These degenerative changes interact in complex ways with progression of Alzheimer's disease and the atrophy induced by it [18]. See Fig. 1 for MRI scans of brains showing degeneration in brain tissue. Thus, understanding these changes and their different manifestations will be crucial to the prevention and management of the diseases.

Aging related symptoms – atrophied regions, lesions etc. – affect connectivity in the brain; and aberrations in connectivity are also closely related to neural disorders including Alzheimer's, Autism Spectrum Disorder and many others [23].

© Springer Nature Switzerland AG 2021
M. Reyes et al. (Eds.): iMIMIC 2021/TDA4MedicalData 2021, LNCS 12929, pp. 119–128, 2021.
https://doi.org/10.1007/978-3-030-87444-5_12

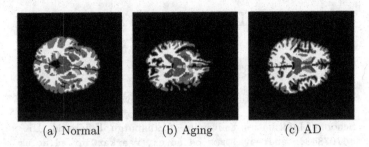

(a) Normal   (b) Aging   (c) AD

Fig. 1. Topological changes with aging and Alzheimer's, showing the effects in gray matter (red) and white matter (white). Both aging and AD result in the loss of white matter volume and a thinning of the gray matter. Gray matter of the Aging brain breaks down into smaller connected components. (Color figure online)

In this paper, we study the changes in *connectivity* of the brain by observing the changes in its topology. Using data analysis on structural MRI scans, we identify signatures of deterioration in connectivity, and develop techniques to isolate the signs of Alzheimer's from ordinary aging.

Our Contributions. We take the approach of persistent homology – which is a way to describe topology at multiple scales or function values. Persistent homology produces artifacts called Betti Curves [7] that count the number of disconnections and holes. Our approach computes persistent homology separately for different functions for white and gray matter, and we find that the resultant Betti curves make accurate predictors of the chronological age. These ideas are described in Sect. 4.1.

Since changes in connectivity are common features of aging as well as of Alzheimer's, we develop a method to factor out the topological signature of aging to identify cases of Alzheimer's (Sect. 4.2). This method works using the relation between chronological age and Betti curves. It first predicts a Betti curve given the chronological age of the patient. Next, it computes the Betti curve of the MRI scan, and finds the *residual Betti curve* – the difference between the computed and predicted curve. This residual curve is found to produce significant gains in identification of Alzheimer's disease from MRI scans.

Experimental results (Sect. 5) show that topological features are good predictors of both age and Alzheimer's, and the newly defined residual Betti curves factor out the aging affect from Alzheimer's fairly successfully. In particular, we obtain an R^2 score of 66% and Mean Absolute Error (MAE) of 4.47 years for the age prediction task (Sect. 5.1) and $F1$ score of 0.62 and 0.74 balanced accuracy for detecting Alzheimer's. These results are comparable to recent CNN based methods [30] (Sect. 5.2). Thus, our results suggest that the topological changes are an important descriptor of the structural changes in the brain and should be investigated in greater detail for understanding of aging and Alzheimer's.

2   Related Works

A multitude of automated methods to estimate brain age [8,14] and detect Alzheimer's have been suggested in the literature in recent years, with their success highly dependent on the set of participants, image preprocessing procedures, cross-validation procedure, reported evaluation metric, number and modality of brain scans [30]. For detection of AD, they generally fall into two main categories. The first one is those that classify AD based on biomarkers (e.g. [18,24]), where the biomarkers have been generally based around describing brain atrophy [4]. The second one is for studies based on deep learning and convolutional neural networks (CNNs) [1,3,22].

The studies in the first group mainly focus on measuring brain atrophy from a structural MRI, which is an important biomarker for determining the status and strength of the neurodegeneration from AD [29]. However, the existing methods are known to be incomplete. Though the deep learning approaches from the second group show promising results, there are challenges in terms of generalisation [2,26], bias and data leakage [30]. A recent summary of CNN-based methods and a number of possible issues can be found in [30].

Topological data analysis (TDA) has been used to study other static anatomical data (e.g. retina images [5,17]) as well as dynamic data such as functional MRI (fMRI) scans [27]. fMRI represents activity in different parts of the brain by detecting blood flow. In contrast, MRI (or structural MRI) produces a static anatomical image and not of the temporary activity. To the best of our knowledge, the current paper is the first TDA-based analysis of AD on MRI data of the brain, showing that Topology is a significant feature of aging and AD.

3   Preliminaries

3.1   Homology and Betti Curves

Betti numbers are *topological invariants* that count numbers of holes in each dimension. The dimension n Betti number is written as β_n. For example, β_0 counts the number of connected components, β_1 is the number of one-dimensional or "circular" holes and β_2 counts the number of enclosed spaces or voids. We present here only a basic description needed for our exposition. For a detailed mathematical treatment, we refer the reader to various excellent texts on algebraic topology [9,19].

The Betti number is sensitive to noise and small perturbations in data. Topological data analysis (TDA) is made robust through the use of topological persistence [6,31] – working off the assumption that structures which persist over multiple scales of data, or multiple values of a relevant *filtration function*, are important. In order to use them in our analysis, starting with an image I, we build a representation of the data as cubical complexes F_r, according to the sub-level sets of some carefully selected filtration function $f : I \rightarrow \mathbb{R}$. All pixels which when used to evaluate f fall below some threshold parameter r are included in the cubical complex, and, by increasing r, we obtain a sequence of nested cubical complexes $F_{r_0} \subset F_{r_1} \subset F_{r_2} \subset \ldots$. The *Betti curve* [7] is the sequence consisting of the Betti numbers of these complexes: $B_n(I, r) := \beta_n(F_r)$.

3.2 Dataset

The dataset used in this study is a publicly available dataset of MRI scans OASIS-3 [21]. It consists of scans of 1098 individuals, taken over a period of 10 years as part of a longitudinal neuroimaging study. We have used a portion of the dataset consisting of 888 scans, where 733 are healthy and 155 with AD. Table 1 shows the gender and age distribution. The dataset was chosen for the availability of precomputed FreeSurfer [10] files which include skull-stripped scans, brain segmentations [12], cortical thickness measurements [11] and 3D reconstructions. This allows for a much more streamlined analysis of the images, and means that we can relate the results directly to the precomputed measurements of brain volumes and thicknesses. These measurements form the feature sets for the *Baseline model* used in experiments.

Table 1. Demographic data of the subset used in experiments.

	Healthy	AD
Age (mean ± SD)	66.9 ± 9.3	74.6 ± 8.1
Gender (M/F)	258/475	74/81

4   Algorithms and Methods

4.1   Betti Curve Based Features on MRI Data

To compute the Betti curves of MRI scans, we need an appropriate filtration function. Given the different qualitative properties we seek to capture between gray and white matter, we isolate the two regions and compute the persistent homology using two different filtration parameters.

The gray matter is processed using density filtration [16]. This function smooths out the image by assigning to each pixel the number of non-zero neighbours within a 3 pixel radius. For white matter, we wish to detect discolourations associated with lesions, thus the filtration is performed directly on the pixel intensity values.

Persistence diagrams are computed on filtration values. These are then normalized to $[0, 1]$ and used to compute the Betti curves (E.g. see Fig. 2). Additionally, Freesurfer's preprocessing procedure produces point clouds outlining a 3D reconstruction of the outer surface of the brain. We compute the Betti curves for these point clouds from Vietoris-Rips filtration [6]. The persistent homology computation produces Betti curves for dimensions 0, 1 and 2 using 100 filtration values in each dimension, resulting in three 100-dimensional vectors.

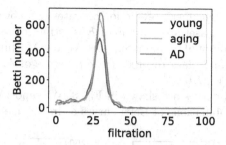

Fig. 2. Example of a dimension 1 Betti curve computed from the gray matter. The age of subjects are 56 for young, 65 for AD and 83 for aging.

For the machine learning algorithms, we concatenate these curves from different dimensions into a single vector, and call this the Betti curve $B(s)$ for subject s. Concatenation of curves from different combination of regions of the brain (gray matter, white matter and surface reconstruction) produces different instances of $B(s)$.

4.2   Aging vs Alzheimer's

We first develop techniques to correlate age with the Betti curve features, and then compute *Residual Betti Curves* that represent Betti Curves modulo the effect of age.

Predicting Age from Betti Curves. The age prediction model is built using a random forest regressor [20] on the Betti curves. Several models are trained for each set of the Betti curves (white matter, gray matter and point cloud) independently. We train an additional model consisting of the various Betti curves concatenated together. We compare the performance of all these models to a baseline model consisting of a random forest regressor trained on the FreeSurfer volumetric statistics.

Residual Betti Curves: Factoring Out the Effect of Age. Next, we consider the inverse problem of predicting the combined Betti curves from the age of a person. For a person of age a, we call this the expected Betti curve $E(a)$. The true combined Betti curve $B(s)$ can be regarded as a sum:

$$B(s) = E(a) + R(s),$$

where we call $R(s)$ the residual Betti curve for subject s of age a. $R(s)$ is the difference between $B(s)$ and $E(a)$ and represents how much the brain structure deviates from the expected (healthy) brain. Values in R will be usually small for healthy subjects and large for AD patients.

The classification model is then trained on the set of residuals to obtain the distribution $P(A|R(s))$ where A is a binary random variable indicating the presence of Alzheimer's.

This training for Alzheimer's prediction operates as follows. Normalized Betti curves for the entire dataset are computed. E is estimated using a linear regression on a healthy subpopulation. A Support Vector Machine (SVM) model is trained on the remaining data to predict the presence of Alzheimer's given the residuals $R(s) = B(s) - E(a)$

Example residual curves are shown in Fig. 3. We report the test set performance of various combinations of Betti curves as in the age prediction model.

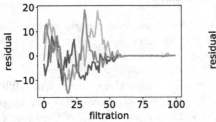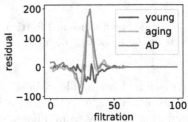

Fig. 3. Comparison of gray matter residual curves in the dimensions 0 and 1. The first dimensional residual (right), shows clearer distinction between the old subject and the AD patient. The age of subjects is 56 for young, 65 for AD and 83 for aging.

4.3   Implementation Details

The source code is available on `GitHub`[1], where more implementation details can be found. Persistent homology was computed with `giotto-tda` library [28]. Baseline comparisons were conducted with a set of standard features. The random forest age prediction model, as well as the SVM and linear regressor models of the AD classification model are instantiated with the default hyperparameters in `scikit-learn`. For the former, sampling procedure is stratified to preserve the distribution of ages in the data, resulting in 666 train curves and 466 test curves. For the latter, the linear regressor is trained on 200 healthy curves and the SVM is trained on 566 curves (100 more curves are added to the remaining 466 by oversampling due to imbalance between subjects with and without AD).

5   Experimental Results

5.1   Brain Age Prediction

Initial experiments explore the utility of Betti curves as predictors of brain age. This is done by training several random forest models on the Betti curves obtained from the various regions, white matter, gray matter and point cloud surface reconstructions. We investigate the performance of the regression when restricted to the Betti curves of a particular region as well its performance when the Betti curves are concatenated together. The *Baseline* model consists of a random forest regressor trained on the FreeSurfer volumetric statistics.

---

[1] https://github.com/ameertg/BrainAgingTDA.

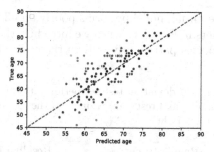

Fig. 4. Predicted versus actual age for healthy subjects.

The correlation between the true and predicted ages can be seen in Fig. 4. For evaluation, we consider the R^2 score on the testing set. This is a common metric used in the evaluation of regression models and describes the variance of the target variable explained by the model. An R^2 score of 1 indicates perfect fit while 0 indicates no correlation between the model output and the target labels.

Table 2 indicates that for raw MRI the combined white and gray matter features performs well with an R^2 score of 0.51, while the addition of point cloud information provides a large increase to an R^2 score of 0.66 and Mean absolute error of 4.47 years.

Table 2. Mean test set model performances (mean and standard deviation) on 5 runs (with random train and test sets) in terms of R^2 score and Mean Absolute Error (MAE) for various feature sets. Based on combined Betti curve features.

Model	R^2 score	MAE
Freesurfer baseline	0.46 ± 0.02	5.43 ± 0.21
White matter	0.43 ± 0.02	5.80 ± 0.22
Gray matter	0.49 ± 0.03	5.56 ± 0.17
White + gray matter	0.50 ± 0.02	5.46 ± 0.24
Surface point cloud	0.53 ± 0.03	5.13 ± 0.32
All combined	$\mathbf{0.65 \pm 0.02}$	$\mathbf{4.47 \pm 0.23}$

5.2   AD Detection with Age Correction

Here we compare two approaches to training an AD classifier. The first method uses the Betti curve obtained from the MRI scan while the second method incorporates information about aging by considering the residual Betti curves. As above, we test both methods on individual Betti curves and also on the concatenated curve which includes the white, gray and point cloud Betti curves. Since the number of healthy and AD patients are different, we compute F1 score and balanced accuracy which are different methods for interpreting unbalanced data.

Table 3 shows that the initial model performs poorly on all feature sets but sees significant improvements in F1 scores when we introduce the residual curve. The residual approach performs particularly well on the gray matter features.

Table 3. Mean and standard deviation test set performance of the SVM classifier on 5 runs (with random training and test sets) trained on the raw Betti curves (left); and on the residual Betti curves (right).

	Raw Betti curves		Residual Betti curves	
	F1 score	Balanced acc	F1 score	Balanced acc
Gray matter	0.00 ± 0.0	0.50 ± 0.0	$\mathbf{0.61 \pm 0.02}$	$\mathbf{0.74 \pm 0.03}$
White matter	0.13 ± 0.01	0.51 ± 0.02	0.50 ± 0.03	0.65 ± 0.04
Surface point cloud	$\mathbf{0.42 \pm 0.02}$	0.76 ± 0.03	0.45 ± 0.03	0.57 ± 0.03
Combined	0.30 ± 0.02	$\mathbf{0.76 \pm 0.02}$	0.47 ± 0.04	0.60 ± 0.03
Baseline	0.36 ± 0.02	0.66 ± 0.02	0.22 ± 0.03	0.47 ± 0.02

6   Discussion

The most striking observation from our experiments is that simple classification on topological features produces a balanced accuracy of 0.76 on Raw Betti Curves and 0.74 on Residual Betti curves, while in comparison, state-of-the-art CNN-based methods (3D subject-level CNN, 3D ROI-based CNN, 3D patch-level CNN, 2D slice-level CNN) achieve average balanced accuracy varying between 0.61 and 0.73 (see Table 6 in [30], last column, row Baseline). Our experiments use about half of the same dataset OASIS-3 [21], and thus the comparison needs some further study. However, it is clear from the results that topology can play a significant role in the study and detection of Alzheimer's disease.

Various other results in our study also point toward specific relevance of topology in the study of the brain. The residual Betti curves on Gray matter were particularly accurate, suggesting the need for further study of the topology of gray matter. Similarly, the topological age prediction was explored here only to the extent it was useful for subsequent detection of AD. Its combination with other techniques such as BrainAge [15] and relation to the Freesurfer and CNN features remain to be explored.

Acknowledgements. RA is supported by the UKRI (grant EP/S02431X/1).

References

1. Aderghal, K., Benois-Pineau, J., Afdel, K., Gwenaëlle, C.: FuseMe: classification of sMRI images by fusion of deep CNNs in 2D+ε projections. In: Proceedings of the 15th International Workshop on Content-Based Multimedia Indexing, pp. 1–7 (2017)
2. Andreeva, R., Fontanella, A., Giarratano, Y., Bernabeu, M.O.: DR detection using optical coherence tomography angiography (OCTA): a transfer learning approach with robustness analysis. In: International Workshop on Ophthalmic Medical Image Analysis, pp. 11–20. Springer, Cham (2020). https://doi.org/10.1007/978-3-030-63419-3_2
3. Bäckström, K., Nazari, M., Gu, I.Y.H., Jakola, A.S.: An efficient 3D deep convolutional network for Alzheimer's disease diagnosis using MR images. In: 2018 IEEE 15th International Symposium on Biomedical Imaging (ISBI 2018), pp. 149–153. IEEE (2018)
4. Beheshti, I., Demirel, H., Initiative, A.D.N., et al.: Feature-ranking-based Alzheimer's disease classification from structural MRI. Magn. Reson. Imaging **34**(3), 252–263 (2016)
5. Beltramo, G., Andreeva, R., Giarratano, Y., Bernabeu, M.O., Sarkar, R., Skraba, P.: Euler characteristic surfaces. arXiv preprint arXiv:2102.08260 (2021)
6. Carlsson, G.: Topology and data. Bull. Am. Math. Soc. **46**(2), 255–308 (2009)
7. Chung, Y.M., Lawson, A.: Persistence curves: a canonical framework for summarizing persistence diagrams. arXiv preprint arXiv:1904.07768 (2019)
8. Cole, J.H., et al.: Brain age predicts mortality. Mol. Psychiatry **23**(5), 1385–1392 (2018)
9. Edelsbrunner, H., Harer, J.: Computational Topology: An Introduction. American Mathematical Society (2010)
10. Fischl, B.: Freesurfer. Neuroimage **62**(2), 774–781 (2012)
11. Fischl, B., Dale, A.M.: Measuring the thickness of the human cerebral cortex from magnetic resonance images. Proc. Nat. Acad. Sci. **97**(20), 11050–11055 (2000)
12. Fischl, B., et al.: Whole brain segmentation: automated labeling of neuroanatomical structures in the human brain. Neuron **33**(3), 341–355 (2002)
13. Fjell, A.M., et al.: High consistency of regional cortical thinning in aging across multiple samples. Cereb. Cortex **19**(9), 2001–2012 (2009)
14. Franke, K., Gaser, C.: Longitudinal changes in individual brainAGE in healthy aging, mild cognitive impairment, and Alzheimer's disease. GeroPsych J. Gerontopsychol. Geriatr. Psychiat. **25**(4), 235 (2012)
15. Franke, K., Gaser, C.: Ten years of brainAGE as a neuroimaging biomarker of brain aging: what insights have we gained? Front. Neurol. **10**, 789 (2019)
16. Garin, A., Tauzin, G.: A topological "reading" lesson: classification of MNIST using TDA. In: 2019 18th IEEE International Conference on Machine Learning and Applications (ICMLA), pp. 1551–1556. IEEE (2019)
17. Giarratano, Y., et al.: A framework for the discovery of retinal biomarkers in optical coherence tomography angiography (OCTA). In: Fu, H., Garvin, M.K., MacGillivray, T., Xu, Y., Zheng, Y. (eds.) OMIA 2020. LNCS, vol. 12069, pp. 155–164. Springer, Cham (2020). https://doi.org/10.1007/978-3-030-63419-3_16
18. Habes, M., et al.: Advanced brain aging: relationship with epidemiologic and genetic risk factors, and overlap with Alzheimer disease atrophy patterns. Transl. Psychiatr. **6**(4), e775–e775 (2016)
19. Hatcher, A.: Algebraic Topology. Cambridge University Press, Cambridge (2002)

20. Ho, T.K.: Random decision forests. In: Proceedings of 3rd International Conference on Document Analysis and Recognition, vol. 1, pp. 278–282. IEEE (1995)
21. LaMontagne, et al.: OASIS-3: longitudinal neuroimaging, clinical, and cognitive dataset for normal aging and Alzheimer disease. MedRxiv (2019)
22. Liu, J., et al.: Applications of deep learning to MRI images: a survey. Big Data Min. Anal. **1**(1), 1–18 (2018)
23. Ouyang, M., Kang, H., Detre, J.A., Roberts, T.P., Huang, H.: Short-range connections in the developmental connectome during typical and atypical brain maturation. Neurosci. Biobehav. Rev. **83**, 109–122 (2017)
24. Papakostas, G.A., Savio, A., Graña, M., Kaburlasos, V.G.: A lattice computing approach to Alzheimer's disease computer assisted diagnosis based on MRI data. Neurocomputing **150**, 37–42 (2015)
25. Patterson, C., et al.: World Alzheimer report 2018 (2018)
26. Qayyum, A., Qadir, J., Bilal, M., Al-Fuqaha, A.: Secure and robust machine learning for healthcare: a survey. IEEE Rev. Biomed. Eng. **14**, 156–180 (2020)
27. Rieck, B., et al.: Uncovering the topology of time-varying fMRI data using cubical persistence. In: Advances in Neural Information Processing Systems, vol. 33 (2020)
28. Tauzin, G., Lupo, U., Tunstall, L., Pérez, J.B., Caorsi, M., Medina-Mardones, A.M., Dassatti, A., Hess, K.: giotto-tda: a topological data analysis toolkit for machine learning and data exploration. J. Mach. Learn. Res. **22**, 39–1 (2021)
29. Vemuri, P., Jack, C.R.: Role of structural MRI in Alzheimer's disease. Alzheimer's Res. Ther. **2**(4), 1–10 (2010)
30. Wen, J., Thibeau-Sutre, E., et al.: Convolutional neural networks for classification of Alzheimer's disease: overview and reproducible evaluation. Med. Image Anal. **63**, 101694 (2020)
31. Zomorodian, A., Carlsson, G.: Computing persistent homology. Discrete Comput. Geom. **33**(2), 249–274 (2005)

Author Index

Printed in the United States
by Baker & Taylor Publisher Services